Green Energy and Technology

Climate change, environmental impact and the limited natural resources urge scientific research and novel technical solutions. The monograph series Green Energy and Technology serves as a publishing platform for scientific and technological approaches to “green”—i.e. environmentally friendly and sustainable—technologies. While a focus lies on energy and power supply, it also covers "green" solutions in industrial engineering and engineering design. Green Energy and Technology addresses researchers, advanced students, technical consultants as well as decision makers in industries and politics. Hence, the level of presentation spans from instructional to highly technical.

Indexed in Scopus.

Indexed in Ei Compendex.

Abdul Sattar Jatoi

Biological Removal of Sulphur from Coal

Past, Current and Future Perspective

Springer

Abdul Sattar Jatoi
Department of Chemical Engineering
Dawood University of Engineering and Technology
Karachi, Pakistan

ISSN 1865-3529 ISSN 1865-3537 (electronic)
Green Energy and Technology
ISBN 978-981-96-8627-8 ISBN 978-981-96-8628-5 (eBook)
https://doi.org/10.1007/978-981-96-8628-5

This Springer imprint is published by the registered company Springer Nature Singapore Pte Ltd.
The registered company address is: 152 Beach Road, #21-01/04 Gateway East, Singapore 189721, Singapore

In the name of Allah, the Most Gracious and the Most Merciful. First of all, I would like to raise my heartfelt gratitude and appreciation to Allah S.W.T for the permission, guidance, wisdom, and blessing for all these years till now, when I have reached this important destination of my journey in life to accomplish my goal. Finally, I would like to present my most heartfelt and warmest appreciation to the great parents and parents-in-law (may ALLAH SWT bless and reward them).

Dr. Abdul Sattar Jatoi

Preface

Coal has been a cornerstone of global energy production for centuries, powering industries and fueling economies. However, the environmental challenges posed by coal utilization, particularly sulfur emissions leading to acid rain and air pollution, have spurred a relentless pursuit of cleaner and more sustainable energy solutions. Among the various methods developed to address these issues, biological desulfurization (BDS) has emerged as a promising, eco-friendly alternative to conventional chemical and physical techniques. This book, *Biological Removal of Sulphur from Coal: Past, Current, and Future Perspective*, presents a comprehensive exploration of BDS, charting its journey from foundational research to modern advancements and future possibilities. It aims to provide a holistic understanding of the processes, mechanisms, and technologies that underpin this innovative approach.

The initial chapters delve into the historical development of BDS, providing insights into the pioneering studies that laid the groundwork for this field. Subsequent sections explore the latest research, including microbial and enzymatic pathways, process optimization, and the integration of BDS with other coal-cleaning technologies. This book also highlights challenges, such as scalability, efficiency, and economic feasibility, offering potential solutions and avenues for future research. This work is intended for a diverse audience, including researchers, engineers, policymakers, and students who are invested in sustainable energy solutions. It seeks to inspire new ideas, foster interdisciplinary collaboration, and underscore the importance of biological innovations in tackling environmental issues associated with coal utilization.

As we stand at the crossroads of an energy transition, this book serves as a reminder that the convergence of biology and engineering holds immense potential for transforming traditional industries. By embracing these innovations, we can move closer to achieving a cleaner, greener future for generations to come.

I hope this book serves as both a valuable resource and a source of inspiration for those dedicated to advancing sustainable technologies in the energy sector.

Karachi, Pakistan

Abdul Sattar Jatoi

Acknowledgment

I thank the Vice Chancellor and higher management of Dawood University of Engineering and Technology and my colleagues in the Department of Chemical Engineering for their support and encouragement.

Abdual Sattar Jatoi

Contents

About the Author

Abdul Sattar Jatoi is an Assistant Professor in the Department of Chemical Engineering at Dawood University of Engineering and Technology, Karachi, Pakistan. He has over 13 years of experience in academics and research. Dr. Jatoi's expertise lies in food bioprocesses and their applications in various food industries, as well as microbial electrochemical systems for environmental and energy applications. He has published more than 140 papers, including 30 book chapters and five books. He is also a peer reviewer for more than 50 scientific journals and has had 70 works published in various international conference proceedings.

His research interests include bio-processing industries, microbial electrochemical system, emerging pollutant treatment via different approaches, sustainable nanomaterials development, wastewater treatment, greenwater treatment technology; separation and purification techniques, and enviro-analytical techniques.

Chapter 1
Coal Occurrence and Utilization for Clean Energy Production

Abstract Coal mining has a history spanning thousands of years. It became particularly important during the Industrial Revolution of the nineteenth and twentieth centuries when it was extensively used to power steam engines, provide heating, and generate electricity. Even today, coal mining remains a key economic activity. Coal produces more energy per mass than wood fuels and is frequently available in places where wood is difficult to come by. Although historically used as a residential fuel, coal is currently mostly employed in industry, particularly in the manufacturing of alloys and smelting, as well as in the creation of electricity. This chapter gives a thorough introduction to coal and its background, history, and chemistry. Hydrogen, oxygen, nitrogen, ash, and sulfur are all present in coal. In terms of coal burning, it produces more ash and sulfur, which are problematic for both the environment and human health.

Keywords Coal · History · Occurrence · Chemistry

1.1 Introduction

When the Earth was covered in massive swampy forests hundreds of millions of years ago, coal was created as a fossil fuel from the leftovers of plants that had lived and perished. It takes millions of years for coal to produce, which is why it is referred to be a nonrenewable energy source. Energy that plants acquired from the sun hundreds of millions of years ago is what fuels coal today. Every live plant stores solar energy [1–3]. Usually, this energy is released when the plants decompose after they die. However, under some circumstances, the decay is stopped, stopping the release of the solar energy that has been stored. Plants that dropped to the bottom of the marsh started to decompose between 100 and 400 million years ago when mud and water were stacked on top [4, 5]. These layers' heat and pressure induced a chemical transformation that over time resulted in coal, a sedimentary rock. Coal, commonly referred to as coal seams, is a combustible sedimentary rock that appears black or brownish-black in color. It primarily consists of carbon, along

A. S. Jatoi, *Biological Removal of Sulphur from Coal*, Green Energy and Technology, https://doi.org/10.1007/978-981-96-8628-5_1

with varying amounts of hydrogen, sulfur, oxygen, and nitrogen. The formation of coal begins with the decomposition of dead plant material into peat, which is subsequently converted into coal through the effects of heat and pressure from deep burial over millions of years [6, 7]. The vast majority of the world's tropical land regions were covered by coal forests, or former wetlands, during the late Carboniferous (Pennsylvanian) and Permian periods. Many large coal deposits, meanwhile, are from the Mesozoic and Cenozoic periods, which are much younger than this. Fuel is the main use for coal. Although coal has been utilized for thousands of years, there has been little use of it before the Industrial Revolution. The development of the steam engine led to an increase in coal usage. In 2016, coal accounted for nearly a quarter of the world's primary energy and two-fifths of the electricity produced. Some industrial processes, including those used to make iron and steel, burn coal [8].Coal consumption and mining result in several early deaths and illnesses. As the greatest producer of anthropogenic carbon dioxide, or 14.4 gigatonnes (Gt) in 201, accounting for 40% of all fossil fuel emissions and more than 25% of all global greenhouse gas emissions, the coal sector harms the environment, notably via climate change. Many nations have decreased or stopped using coal as a source of energy as part of the global energy transition, and the UN Secretary General has urged states to cease constructing new coal plants by the year 2020. In order to achieve the Paris Agreement's goal of keeping global warming well below 2 °C (3.6 °F), coal consumption must more than halve between 2020 and 2030 [9, 10]. China is both the biggest user and importer of coal. China mines about half of the coal produced worldwide, with India mining the remaining 10%. About a third of the coal exported worldwide comes from Australia, followed by Indonesia and Russia [11, 12]. This chapter cover all the aspects related to coal regarding their occurrence, history, and chemistry.

1.2 Coal History

When rocks are deposited on top of peat, pressure from the rocks causes coal to develop. The Pennsylvanian Epoch (Carboniferous Period) saw the formation of a large portion of the world's first coal from the remnants of plants that had existed and perished in tropical swamps millions of years before. Coal is a sedimentary rock in this form, but anthracite coal, with its hard, shiny luster, is a metamorphic rock because it was subjected to higher pressures and temperatures. The best coal is often anthracite coal [13, 14]. China is the main producer of anthracite, accounting for more than 75% of the world's supply. Given the blatantly glaring lack of technology or understanding in comparison to what is available now, the history of humankind's connection with coal is necessarily hazy [15]. However, the Aztecs were among of the first people in the Americas to utilize coal, and they did so for fuel. In order to fully use Roman Britain's coalfields, the Romans transformed Britain into a coal hotspot in the second century AD. The remains of coal storage have been uncovered during archaeological digs at several forts along the renowned

Hadrian's Wall during the succeeding centuries. Longovicium, a neighboring fort, has proof that the Romans established a smelting business in Northern England. Greek scientist Theophrastus described coal in his book *On Stones* before the Romans were interested in the early coal business. The earthen elements known as coals are among those that are excavated because they are valuable; when lit on fire, they burn like coal. They may be seen in Liguria and in Elis as one travels up the mountain route to Olympia. Metalworkers utilize them. Theophrastus' writings were among the first, as far as historians and archaeologists are aware, to mention the use of coal in metallurgy. However, it is believed that the Chinese may have started coal surface mining as early as 3490 BC. Drift mining, which often involves a horizontal tunnel that follows the coal bed or an ore vein, for example, bell pit mining, and shaft mining, which frequently incorporates the "room and pillar" operating approach, are some of the early types of coal mining. Mining activities 2000 years ago hardly ever exhausted a mine's bulk of resources. It is unclear if this was because of operational waste or a reluctance to symbolically "empty the lake." Following the fall of the Roman Empire, coal in Britain remained a much less common resource until the Industrial Revolution, whose zenith depended on the widespread availability of coal to power daring steam engines, heat buildings, and, soon after, start being used to create electricity. Due to the employment of coal-fed steam engines for rail and steamship transportation, the Victorian age brought a large expansion in international commerce. In terms of announced statistics, the United Kingdom and the United States of America were in the forefront of coal production around the year 1905. At this time, the United Kingdom was in charge of around 236.1 million short tonnes, whereas America was in charge of 350.8 million. Despite being widely accessible, coal's ascent to prominence throughout the Industrial Revolution was due to several factors. Additionally, it was a cheap resource that promised to produce more energy than competing materials like wood. Landscapes have changed as a result of the drastic changes in weather and climate that have occurred since the Pennsylvanian Epoch, 298.9 million years ago. Nowadays, it's common to find areas with readily accessible coal and wood in separate locations.

Coal is primarily used for energy and heat, having been used since the early AD period of Roman history in the Rhineland to smelt iron ore, likely for weapons, wagon wheels, and armor, as well as in hypocausts to heat public and military baths, through to powering the British Empire's trade expeditions. According to recent scientific projections, coal provided around 25% of the world's energy in 2010; by 2050, this is anticipated to increase to 33%. Nearly 70% of the world's steel manufacturing depends on the combustion of coal. Over 255 million short tonnes of coal are produced in South Africa, one of the many countries where WRS has considerable mining experience. This country also accounts for 92% of all coal used on the African continent, much of it used for electricity. China does generate the most coal, not only anthracite coal. China has produced between 2.9 and 3.9 billion short tonnes annually during the last 10 years on average. China's primary energy source is coal, which employs around five million people annually. The Chinese mining colossus is not going to give up anytime soon, as was highlighted in several of our other recent stories on mining.

Coal mining has a rich history, with early mines traced back to the Roman Empire, ancient China, and other early civilizations. Its importance grew significantly during the Industrial Revolution of the nineteenth and twentieth centuries, when it was primarily used to power steam engines, heat buildings, and generate electricity. While coal mining remains an important economic activity today, its prominence is declining due to its significant contribution to climate change and other environmental issues, leading to reduced demand and, in some regions, a peak in coal usage.

Coal produces more energy per unit mass than wood and is often accessible in areas where wood is scarce. Although historically used as a residential fuel, coal is now predominantly utilized in industrial applications, particularly in alloy production, smelting, and electricity generation. From the eighteenth century until the 1950s, coal was the leading source of primary energy for manufacturing and transportation in industrialized regions. Large-scale coal mining developed during the Industrial Revolution, and coal continues to be a major energy source today [1]. Nowadays, coal is also extensively mined using open pit techniques wherever the coal layers are close to the surface or relatively shallow. The principal underground coal mining methods were created in Britain starting in the late eighteenth century, and advancements in the nineteenth and early twentieth centuries drove further development [1]. But starting in the 1860s, oil and gas were widely employed as substitutes.

By the late twentieth century, oil, natural gas, electricity derived from oil, gas, nuclear power, or renewable energy sources had largely displaced coal in home, industrial, and transportation use. More over a fourth of the world's energy was generated by coal in 2010 [2].

Coal mining has also been a political and social problem since 1890. In the twentieth century, coal miners' labor and trade organizations gained strength, and often, the miners served as leaders of the left or socialist movements (as in Britain, Germany, Poland, Japan, Chile, Canada, and the United States). Environmental concerns have become more significant since 1970, including the health of workers, the degradation of the environment caused by strip mining and mountaintop removal, air pollution, and the impact of coal burning on global warming.

Emissions from burning coal may contaminate the atmosphere. Carbon dioxide, a greenhouse gas, is also produced. A substance called sulfur may also be released when coal is burned. When sulfur and oxygen combine to make sulfur dioxide, a chemical that may harm water and trees when combined with precipitation to form acid rain is created. Companies that mine coal aim for low-sulfur coal. They put forth a lot of effort to purge the coal of sulfur and other impurities. Scrubbers are devices that power stations install to remove the majority of sulfur from coal smoke so that it doesn't enter the atmosphere. Previously, landfills were used to dispose of other byproducts, such as the ash left behind after burning coal. Currently, they are used to create ocean reefs that serve as animal habitats, roadways, cement, and cement.

1.3 Coal Occurrence

Coal serves as a widely used source of energy and chemicals. Large sedimentary basins containing rocks of Carboniferous age and younger are present on nearly every continent, including Antarctica [16, 17]. However, the terrestrial plants necessary for coal formation became widespread only during the Carboniferous period (approximately 358.9 to 298.9 million years ago). As a result of climate change and tectonic plate movements, which often shifted ancient landmasses through subtropical and tropical regions, significant coal deposits are now found in areas that currently experience arctic or subarctic climates, such as Alaska and Siberia. Conversely, regions like Greenland and much of northern Canada lack substantial coal deposits because their rocks predate the Carboniferous period, during which the rich terrestrial vegetation needed to form significant coal reserves did not exist. These areas, known as continental shields, are largely devoid of coal reserves.

Estimating global coal resources and reserves is a complex task (Fig. 1.1). In addition to the lack of precise data for specific countries, two broader challenges make such estimates difficult and somewhat arbitrary. The first challenge lies in varying definitions of terms such as "proven reserves" (quantities recoverable under current operational and economic conditions) and "geological resources" (the total quantity of coal present, irrespective of recoverability). Proven reserves should

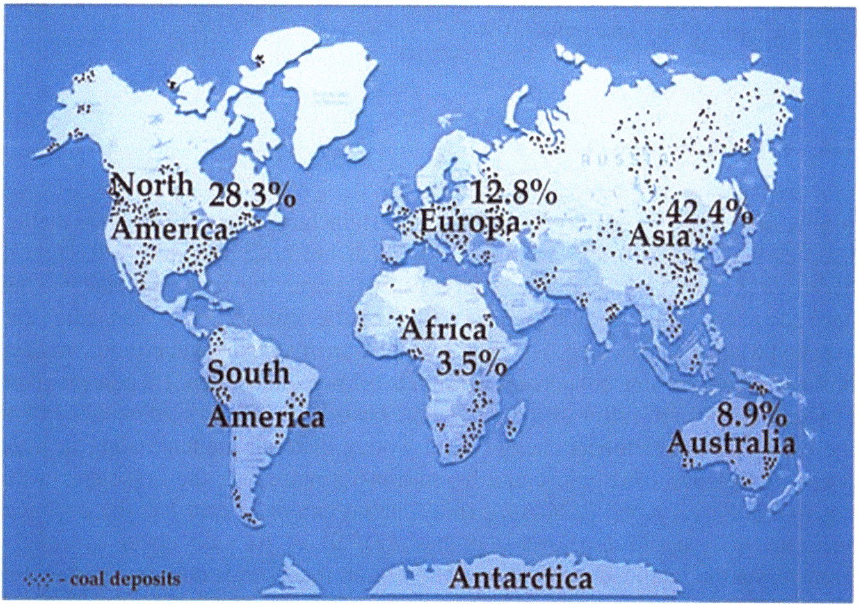

Fig. 1.1 Coal occurrence in the world [18]

ideally represent a reliable estimate of the coal that can be extracted under existing conditions. Economically mineable coal beds typically need to be at least 0.6 meters (2 feet) thick and located no deeper than 2000 meters (6600 feet) below the surface. However, these parameters can vary based on factors like coal quality, market demand, and the feasibility of surface or underground mining. Advances in mining technology may also increase the recoverable coal fraction compared to what is currently inaccessible. For instance, traditional underground mining methods recover only about half of the coal in place, as large coal pillars are left behind to support the overlying rock. In contrast, longwall mining, which employs machinery to extract long, parallel coal seams, can recover nearly all of the coal.

The second challenge involves the consumption rate of coal when estimating reserves. Rather than focusing solely on the total quantity of coal resources, it may be more critical to assess how long coal reserves will last. At current consumption rates, global coal reserves are projected to last between 300 and 500 years. Beyond these reserves, Earth contains a vast amount of additional coal resources that are currently unrecoverable. These "geologic resources" are even harder to estimate but are believed to be up to 15 times greater than proven reserves. Proven coal reserves are often expressed in terms of millions of tonnes of coal equivalent (MTCE). One tonne of coal equivalent represents one metric tonne (2205 pounds) of coal with a heating value of 29.3 megajoules per kilogram (12,600 British thermal units per pound). Based on these measures, the United States appears to have the largest recoverable coal reserves. Approximately 75% of the world's recoverable coal is controlled by five countries: the United States (22%), Russia (15%), Australia (14%), China (13%), and India (10%).

1.4 Coal Chemistry

A proximate study of coal's composition (which includes moisture, volatile matter, fixed carbon, and ash) and an ultimate analysis are both provided (ash, carbon, hydrogen, nitrogen, oxygen, and sulfur) [19, 20]. Except for some methane that has been adsorbed, "volatile matter" refers to the volatile substances that are created and expelled when coal is heated. The final composition of a typical bituminous coal can be 84.4 percent carbon, 5.4 percent hydrogen, 6.7 percent oxygen, 1.7 percent nitrogen, and 1.8 percent sulfur on a dry, ash-free basis [21]. During coalification, all of the original plant elements change, with woody material predominating in most coals. Elimination of cellulose and lignin transformation are the first steps in the transformation of wood [22]. As a result, sub-bituminous coal has the chemical structure of a lignin precursor that has lost all of its side chain hydroxyls. Transformation that leads to the production of aromatic rings during the evolution of sub-bituminous coal into bituminous coal causes an increase in aromaticity. A very fragrant structural characteristic distinguishes anthracite from semi-anthracite.

Small aromatic stacks gather into clusters at the beginning of the semi-anthracite rank when an abrupt molecular orientation occurs [23]. On the road to graphite, interlayer spacing eventually shrinks toward anthracitic rank. The composition of coal gradually changes as a consequence of these biological material transitions, both chemically and structurally. The most noticeable aspects of elemental chemistry are increases in carbon concentration and reductions in oxygen content [24]. Each maceral group has its unique development route while following common patterns in elemental composition because of variances in their initial chemical makeup. Generally speaking, the aromaticity of macerals rises from liptinite to vitrinite to inertinite [25]. As a result, they react differentially to the processes of coalification, as shown in Fig. 1.2 for the carbon content assessed in situ by electron microprobe on individual macrophytes. With successive coalification, functional group abundance and distribution also change and are particular to certain types of minerals (macerals). Analyses of maceral concentrates obtained through density gradient centrifugation (DGC), along with in situ techniques such as electron microprobe and micro-FTIR (micro-Fourier transform infrared spectroscopy), reveal

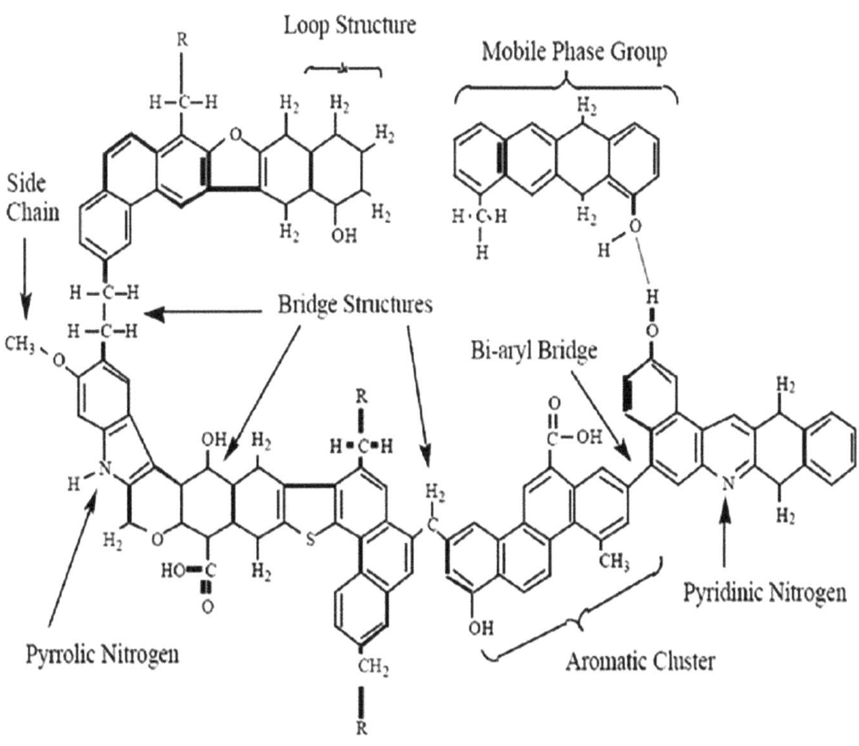

Fig. 1.2 Coal structure [26]

significant chemical differences both between and within macerals of iso-rank coals. Additionally, the chemistry of macerals changes with coal rank. Figure 1.2 presents examples of recent micro-FTIR mapping of maceral surfaces in high-volatile A bituminous coals, highlighting notable variations in the abundance of aliphatic stretching C-H groups. These differences are evident not only within relatively homogeneous macerals like vitrinite and semifusinite but also between distinct macerals, such as vitrinite and sporinite. Such substantial variations within individual macerals can significantly influence the behavior of coal in response to changes in its surface chemistry.

1.4.1 Hydrophobicity of Coal

Higher-rank coals are said to be hydrophobic by nature. This is a consequence of the chemistry and hydrocarbon structure of those substances. Due to the presence of oxygen functional groups and the kind of hydrocarbons in coal's makeup, its surface qualities determine how water is attracted to it and how hydrophilic or hydrophobic it is [27]. Oxygen functional groups within the coal structure serve as sites for water adsorption on the coal surface. In lower-rank coals, these groups, such as hydroxyl, carbonyl, and carboxyl, may originate from the initial coal structure or form through oxidation. The amount and composition of these oxygen functional groups vary with the coal's rank. As shown in Fig. 1.3, the variation in the quantity and type of oxygen functional groups is commonly used to correlate changes in coal surface properties with its rank [28, 29]. It is well known that coal's hydrophobicity depends on rank. However, hydrophobicity is principally influenced by the surface heterogeneity resulting from the petrographic composition (maceral composition) in conjunction with minerals related to coal for a specific rank of coal and regulated amount of oxidation. Due to the variances in the chemical makeup of minerals, their surface qualities also differ [30].

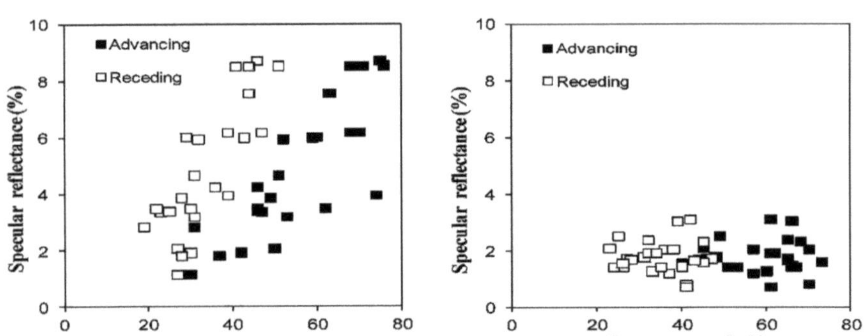

Fig. 1.3 Hydrophobicity of coal [31]

1.5 Sulfur in Coal

In general, there are two main categories of sulfur-containing chemicals found in fossil fuels (crude oil and coal): inorganic sulfur and organic sulfur. Sulfides and sulfates make up the majority of inorganic sulfur. Pyrite (FeS_2), chalcopyrite ($CuFeS_2$), sphalerite (ZnS), galena (PbS), arsenopyrite (FeAsS), molybdenite (MoS_2), covellite (CuS), argentite (Ag2S), cinnabar (HgS), and other minerals are mostly sulfide minerals. Barite ($BaSO_4$), gypsum (CaSO4.2H_2O), anhydrite ($CaSO_4$), and other iron sulfates are among the sulfate minerals [32, 33]. The primary inorganic sulfur component found in coal matrix is often pyrite. In contrast to pyritic or inorganic sulfur, organic sulfur is covalently bonded within the massive, complex structures of coal and is difficult to extract physically or chemically. There are four different types of organic sulfur found in coal, including aliphatic, aromatic, and heterocyclic forms [33]:

The majority of sulfur in coal samples is inorganic (pyritic), although sometimes organic sulfur makes up more than 50% (and even up to 70%) of the total sulfur in coal. The inclusions of pyritic sulfur do not form a fundamental element of the coal structure; rather, they are present. Pyrite is found in the form of particles, sometimes in bands, other times agglomerated into balls or nodules, although it is typically extremely finely dispersed. Several physical and chemical methods, including heavy media separation, magnetic separation, leaching, selective agglomeration, and flotation, may be used to extract pyrite sulfur efficiently [34]. Thiols, sulfides, disulfides, and thiophenes are the four types of organic sulfur found in coal. Physical techniques cannot be used to remove organic sulfur from coal because it is covalently bonded to the coal's structure and is also finely dispersed. After carbon and hydrogen, sulfur is regarded as the third most prevalent element in petroleum. Higher-density crude oils may hold more sulfur compounds. Higher quantities of sulfur compounds are also present in distillation fractions with higher boiling points [35]. Similar to coal, crude oil naturally contains both organic and inorganic molecules that include sulfur. Elemental sulfur, H_2S, and pyrite are examples of inorganic substances that include sulfur. They may be dissolved or suspended. Crude oil often contains aromatic or saturated forms of thiols, sulfides, and heterocycles as organic sulfur compounds [36]. Due to their higher boiling points (more than 200 °C) and the difficulty in removing them from intermediate distillates such atmospheric tower outlet streams, aromatic chemicals like dibenzothiophene (DBT) or its derivatives are of particular significance among them. The typical thiophenic chemicals found in diesel include benzothiophene (BT), non-, single-, and di-substituted benzothiophenes (BP >219 °C) [35]. Some sulfur organic compounds are regarded as refractory. The majority of these chemicals are stable aromatic sulfur-containing compounds that need a more intrusive desulfurization process in order to remove their sulfur atom. Thiophenic compounds, such as DBT derivatives with 4 and/or 6 alkyl substitution groups, are responsible for the refractory component of distillate/diesel fuels [37, 38] (Fig. 1.4).

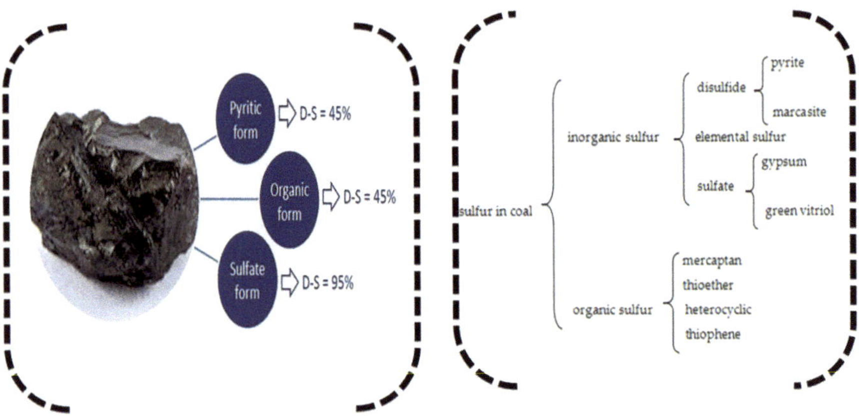

Fig. 1.4 Sulfur forms in coal [38] and their classification [39]

1.6 Challenges Associated with High-Sulfur Coal

Due to a number of technical and environmental issues with the use of high-sulfur fuels, the applications outlined in the previous section call for the use of low sulfur fossil fuels. By burning them, fossil fuels contribute to pollution. Many waste products, including particulates and different gases like sulfur dioxide, nitrogen oxides, and volatile organic compounds, are created as a result of impurities in such fuels [40]. These waste materials might harm both our environment and other creatures, including people. Low pH fogs or acid rain may result from sulfur dioxide (SO_2) interacting with airborne water vapors. Historical structures and other open installations may erode more quickly as a result of the acid that is created in this manner. It may spread to the soil, harm vegetation, lower the pH of lakes with inadequate buffer capacities, and affect marine life (EPA-US, 2006). Additionally, the generation of sulfate aerosols may be triggered by SO_2. The aerosol particles, which may be transferred into the lungs and may result in respiratory disorders, have an average diameter of 2.5 m. Governments and other regulatory bodies from all around the globe have identified the issues related to these emissions and are making specific efforts to decrease them via law. Limits on sulfur emissions from power plants (compensated by low-sulfur fuels, post-combustion cleaning) and the implementation of more rigorous limits on the quantity of sulfur permitted in transportation fuels are two examples of these rules. As a consequence, many regulatory bodies are making efforts to totally remove sulfur from these fuels. Sulfur and nitrogen oxides are released into the environment after the combustion of coal that contains too much sulfur and other minerals, together with a significant quantity of ash or clinker. Clinker jams furnace grate openings, decreasing airflow and, in extreme circumstances, potentially leading to furnace shutdown. In areas where large-scale coal burning takes place, either coal with a total sulfur content of less than 1.5% must be used, or measures must be used to scrub sulfur dioxide out of the stack gas

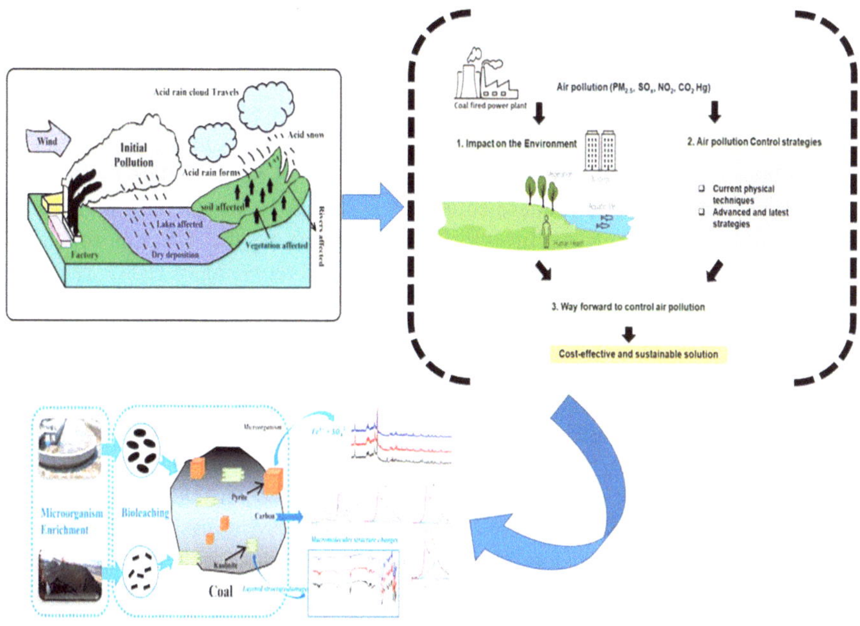

Fig. 1.5 Challenges associated with high sulfur coal [41–43]

before it enters the sky. The coal's sulfur content may contaminate the process catalysts during liquefaction, preventing the reaction from occurring. Similar to this, sulfur compounds in coal may prevent microorganisms engaged in coal biosolubilization from growing. When coal is converted into coke via the coking process, a significant portion of its sulfur content is transferred to the coke, which when employed in metallurgical processes results in brittle steel. As a result, this method also requires the use of low sulfur and ash content coal. By using clean coal with minimal sulfur content, environmental pollution and other issues related to coal use may be reduced (Fig. 1.5).

1.7 The Current Legislative Regulations Regarding Sulfur Emission from Coal Combustion in Different Countries

The direct limitation of SO_3/H_2SO_4 aerosol emissions from utility boilers is presently not covered by any laws in the United States. However, as will be covered subsequently, certain other regulatory initiatives could call for SO_3 regulation. German emission regulations set a daily average sulfur dioxide (SO_2) and sulfur trioxide (SO_3) emission concentration limit of 50 mg/Nm^3 and a 30-min average of 200 mg/Nm^3. There isn't a distinct SO_3-only emission limit. In the United States, H_2SO_4 is classified as a hazardous air pollutant (HAP) under Title III of the Clean

Air Act Amendments of 1990, and SO_3 emissions are included in limitations on opacity and particle matter (PM) emissions. The mass emission limitations of 10 t (9.09 tonnes) per year for any one HAP or 25 t (22.7 tonnes) per year for all HAPs combined imposed for other industrial sources are not applicable to electric utility generating stations under Title III. Therefore, despite the fact that coal-fired utility generating stations produce considerable amounts of H_2SO_4—a classified HAP—there is no legal duty to limit these emissions as HAPs. However, they are required to submit a Toxics Release Inventory (TRI) report to the US Environmental Protection Agency (EPA). Standard measurement and reporting of H_2SO_4 refer to it as comparable SO_3. Primary PM is described by the EPA as particles that are either directly released from the stack as a solid or liquid or released as a vapor but condense or react upon cooling and dilution in the surrounding air to become solid or liquid PM. In the ambient air, secondary PM is made up of particles that arise as a result of chemical processes following dilution and condensation. According to this definition, SO_3 or H_2SO_4 that is in the vapor phase in the furnace or stack and immediately after leaving the stack either reacts to produce SO_4 particles or condenses to produce liquid H_2SO_4 is considered to be a primary PM emission and is consequently subject to PM emission limits. To maintain opacity regulations, plants may be required to manage SO_3 in certain states or municipalities. The creation of a visible plume brought on by the presence of SO_3 may result in plume opacities higher than permitted levels and the need to regulate SO_3 emissions, even when opacity measurements in the stack may be below regulatory limits. For stationary sources, such as coal-fired utility boilers, the majority of states have an opacity limit of 20%; however, many additionally make allowances for greater opacity levels during unit startup. There are a few states that have opacity limitations lower than 20%, including several that have opacity limits as low as 40% and at least one (West Virginia) that has an opacity limit as low as 10%. When in-stack opacity measurement values are lower than those in the downwind plume, a detached plume has formed. While measurements of the in-stack opacity are done using continuous emission monitors, EPA Method 9.9 is applicable to measurements of the downstream plume opacity. Regulation breaches and issues with good neighbor policies might come from the increase in opacity from in-stack measurements to downwind, which is often brought on by the condensation of gases like H_2SO_4 [44].

1.8 Conclusion

This chapter focuses on the coal history, their deposit and occurrences overall the world, as well as their chemistry. Today, coal mining is still a significant economic activity. Coal produces more energy per mass than wood fuels and is frequently available in places where wood is difficult to come by. Although historically used as a residential fuel, coal is currently mostly employed in industry, particularly in the manufacturing of alloys and smelting, as well as in the creation of electricity. In this chapter deposits of coal in the world were discussed and their occurrence overall the world. Coal chemistry was also discussed with respect to their sulfur content, which creates problem for environment as well as for industrial equipment. Furthermore coal can be used via proper cleaning; there are different techniques available for using coal more fruitful way.

References

1. Parker S (2004) Coal. Gareth Stevens Publishing LLLP
2. Schlager N, Weisblatt J (2006) Alternative energy. Thomson Gale
3. Jackson T (2020) Climate change. QEB Publishing
4. Cleal CJ, Thomas BA (2019) Introduction to plant fossils. Cambridge University Press
5. Dufour F (2018) The costs and implications of our demand for energy: a comparative and comprehensive analysis of the available energy resources. In: The Costs and Implications of Our Demand for Energy: A Comparative and Comprehensive Analysis of the Available Energy Resources
6. Speight JG (2020) Synthetic fuels handbook: properties, process, and performance. McGraw-Hill Education
7. Gan, W., Organic petrography and origin of condor coals from Hillsborough Basin, Queensland. 1998
8. McNeill JR, Engelke P (2016) The great acceleration: An environmental history of the Anthropocene since 1945. Harvard University Press
9. Jones H, Muller A (2016) The South African Economy, *1910–90*. Springer
10. Smith C (2018) Coal, steam and ships: engineering, enterprise and empire on the nineteenth-century seas. Cambridge University Press
11. Xiang H, Kuang Y, Li C (2017) Impact of the China–Australia FTA on global coal production and trade. J Policy Model 39(1):65–78
12. Lin B-Q, Liu J-H (2010) Estimating coal production peak and trends of coal imports in China. Energy Policy 38(1):512–519
13. Li J, Hu S (2017) History and future of the coal and coal chemical industry in China. Resour Conserv Recycl 124:13–24
14. Flinn MW, Stoker D (1984) The history of the British coal industry. Volume 2, 1700–1830: the industrial revolution. Clarendon Press, New York
15. Friederich MC, van Leeuwen T (2017) A review of the history of coal exploration, discovery and production in Indonesia: the interplay of legal framework, coal geology and exploration strategy. Int J Coal Geol 178:56–73
16. Armstroff A (2004) Geochemical significance of biomarkers in Paleozoic coals
17. Bend SL (1992) The origin, formation and petrographic composition of coal. Fuel 71(8):851–870
18. Ivanova S et al (2022) An overview of carbon footprint of coal mining to curtail greenhouse gas emissions. Sustainability 14(22):15135
19. Speight JG (2012) The chemistry and technology of coal. CRC Press
20. Gouws SM et al (2021) Co-pyrolysis of coal and raw/torrefied biomass: a review on chemistry, kinetics and implementation. Renew Sust Energ Rev 135:110189
21. Sun L et al (2021) Effects of bluff-body cone angle on turbulence-chemistry interaction behaviors in large-scale semicoke and bituminous coal co-combustion. Fuel Process Technol 221:106915
22. Liang D et al (2019) Evolution of structural and surface chemistry during pyrolysis of Zhundong coal in an entrained-flow bed reactor. J Anal Appl Pyrolysis 140:331–338

23. Medunić G et al (2021) Elevated selenium levels in vegetables, fruits, and wild plants affected by the Raša coal mine water chemistry. Rudarsko-geološko-naftni zbornik (The Mining-Geological-Petroleum Engineering Bulletin) 36(1):1
24. Gómez Rojas OP et al (2020) Petrographic characterization, variations in chemistry, and paleoenvironmental interpretation of colombian coals. Int J Coal Geol 227:103516
25. Bai X et al (2018) Coal production in China: past, present, and future projections. Int Geol Rev 60(5–6):535–547
26. Malumbazo N (2011) Chemical and physical structural studies on two inertinite-rich lump coals. Johannesburg
27. Yu Y et al (2020) New insights into the slime coating caused by montmorillonite in the flotation of coal. J Clean Prod 242:118540
28. Li Y et al (2020) A novel coal tar-based collector for effective flotation cleaning of low rank coal. J Clean Prod 273:123172
29. Li Y et al (2020) Flotation of low-rank coal using sodium oleate and sodium hexametaphosphate. J Clean Prod 261:121216
30. Wang H et al (2020) Enhancement of coal flotation using impact flow conditioning pulp. J Clean Prod 267:122124
31. Ding LP (2009) Investigation of bituminous coal hydrophobicity and its influence on flotation. Energy Fuel 23(11):5536–5543
32. Tang L et al (2020) Effect of removal organic sulfur from coal macromolecular on the properties of high organic sulfur coal. Fuel 259:116264
33. Prayuenyong P (2002) Coal biodesulfurization processes. Songklanakarin J Sci Technol 24(3):493–507
34. Çelik PA et al (2019) The approach of biodesulfurization for clean coal technologies: a review. Int J Environ Sci Technol 16(4):2115–2132
35. Soleimani M, Bassi A, Margaritis A (2007) Biodesulfurization of refractory organic sulfur compounds in fossil fuels. Biotechnol Adv 25(6):570–596
36. Chou C-L (2012) Sulfur in coals: a review of geochemistry and origins. Int J Coal Geol 100:1–13
37. Tang L et al (2018) Feasibility study of reduction removal of thiophene sulfur in coal. Fuel 234:1367–1372
38. da Silva GLR et al. UTILIZAÇÃO DE POLÍMERO EM PILHAS DE CARVÕES E ESTUDO DE DECAIMENTO DE FLUIDEZ E MATÉRIA VOLÁTIL
39. Chen Y, Cao M, Ma C (2019) Review of coal-fired electrification and magnetic separation desulfurization technology. Appl Sci 9(6):1158
40. Li X et al (2018) Clean and efficient utilization of sodium-rich Zhundong coals in China: behaviors of sodium species during thermal conversion processes. Fuel 218:162–173
41. Asif Z et al (2022) Update on air pollution control strategies for coal-fired power plants. Clean Techn Environ Policy 24(8):2329–2347
42. Howladar MF et al (2020) The quality of imported coal and its impact on environmental degradation. Environ Dev Sustain 22(1):251–263
43. Ye J et al (2018) Biodesulfurization of high sulfur fat coal with indigenous and exotic microorganisms. J Clean Prod 197:562–570
44. Lin C-K et al (2018) A global perspective on sulfur oxide controls in coal-fired power plants and cardiovascular disease. Sci Rep 8(1):1–9

Chapter 2
Techniques for Sulfur Removal from Coal

Abstract Through a survey of the literature on desulfurization and a critical evaluation of the different heavy oil desulfurization techniques, strategies were assessed. Hydrodesulfurization, extractive desulfurization, oxidative desulfurization, biodesulfurization, desulfurization by alkylation, chlorinolysis, and the use of supercritical water are among the desulfurization techniques and variants thereof that are studied. For the desulfurization of heavy oil, only a few of these approaches are practical and/or effective. The heavy oil's characteristics, such as its high sulfur content, high viscosity, high boiling point, and the refractory nature of the sulfur compounds, are mostly to blame for this. Autoxidation followed by thermal breakdown of the oxidized heavy oil is the method that has the highest potential to revolutionize the desulfurization of heavy oil. Additionally, there is potential for using autoxidation in concert with hydrodesulfurization, biodesulfurization, and oxidative sulfur reduction.

Keywords Desulfurization techniques · Hydrodesulfurization · Oxidative desulfurization · Biodesulfurization

2.1 Introduction

The most prevalent fossil fuel in the planet is coal [1]. However, due to the release of sulfur dioxide, burning coal with a high sulfur content seriously harms the environment and results in acid rain [2]. The three primary types of sulfur found in coal are pyritic, organic, and sulfate sulfur [3]. Pyritic sulfur is found in coal as mineral matter, whereas organic sulfur is a crucial component of the coal matrix [3]. The organic or inorganic sulfur portions of coal may be removed using a variety of physical-chemical techniques, such as flotation, chemical oxidation, and chemical reduction [4]. However, when these sulfides are finely dispersed in coal matrix, these physical methods are unable to remove the pyrite. Additionally, although organic and inorganic sulfur can be removed from coal using chemical methods,

A. S. Jatoi, *Biological Removal of Sulphur from Coal*, Green Energy and Technology, https://doi.org/10.1007/978-981-96-8628-5_2

this will also result in the loss of partial combustible matter and hazardous secondary products [5]. A safe alternative method to remove sulfur from coals is biodesulfurization, in which microbes catalyze a biochemical reaction in an aqueous medium that causes the sulfur content to oxidize and dissolve into sulfate [6]. Mesophilic microbes, primarily *A. ferrooxidans* and *A. thiooxidans*, have been used frequently to remove pyrite from coals among them. However, only a small amount of literature from the past 10 years claimed that sulfur was removed from coal using thermophilic microbes [7]. *A. caldus* has recently been identified as the main sulfur-oxidizing thermophilic bacterium responsible for the majority of the bioleaching pyrite at 45 °C. As far as we are aware, it has never been used to the biodesulfurization of coal [7, 8]. It is well known that the attachment of cells to mineral surfaces has been linked to the oxidation of sulfide minerals, which depends on both the biochemical characteristics of the cells and the interfacial process between the cells and mineral surface [9]. Furthermore, it has been documented that the growth conditions have a significant impact on the surface characteristics of bacterial cells. In order to better understand the interfacial interaction between cells and mineral surfaces, it is helpful to have a better understanding of the surface characteristics of functional bacteria under various conditions [10, 11]. Numerous studies have been conducted on the bio-oxidation of pyrite, and a potential mechanism has been put forward. This mechanism calls for a stepwise oxidation to produce surface-bound thiosulfate, which is subsequently oxidized through tetrathionate and other polythionates to produce sulfate [12]. Numerous researchers concurred that the effectiveness of sulfur removal may be impacted by the precipitation of mineral phases during the pyrite oxidation process [13]. The sulfur-rich layers may also deposit onto the surface of pyrite in addition to the jarosite that has been described, inhibiting the interaction of microorganisms and impeding the leaching process.

2.2 History and Current Status

Globally, coal has a sulfur content of 1–3 weight percent. All gases and particles released into the atmosphere as a result of chemical reactions and combustion processes, including sulfur emissions from coal, are a global issue. The sulfur in coal reacts with the oxygen during combustion to produce SO_2 and SO_3, which are then expelled down the chimney into the atmosphere. These substances eventually create H_2SO_4 when they interact with atmospheric moisture and fall to Earth's surface as acid rain. Acid rain is significant for the ecosystem and for the harm it does to human life. The first impact of acid rain is seen in plants. The effects of acid rain include leaf spots, leaf dryness, leaf and fruit spills, the beginning of plant development, and harm to forests and agricultural crops as a consequence of plant mortality. Additionally, there is a correlation between rising SO_2 emissions and human respiratory and skin disorders. In addition to these effects, high sulfur content in coal raises processing and maintenance costs for customers by, for example, producing corrosion and low-grade steel in the steel sector. Due to this, it is essential to lower the concentration of gases like SO_2 in order to preserve the environment, safeguard human health, and save industrial costs. Some nations have acknowledged this issue, began to focus on reducing SO_2 emissions, and started to enforce sanctioning this matter. Additionally, the nations of the globe participate in discussions for international accords, particularly those that address concerns like climate change and the effects of greenhouse gases. And this would certainly make the people aware of how fossil fuels are produced. However, coal continues to play a significant role in the global energy supply. Coal is used to generate 40% of the world's electricity. Because of this, using coal without using clean coal technologies (CCTs) will result in harsher legal consequences. In the near future, it will increase public awareness and obstruct several industrial sectors, including the production of iron and steel and energy from thermal power plants. As a result, studies focusing on clean coal technology in the energy industry are being carried out globally. On the other hand, worldwide applications and choices to safeguard the ecological equilibrium for nations with subpar coal deposits, like Turkey, preclude these reserves from being appraised in their entirety. Without the adoption of CCT, these nations increasingly rely on other nations, notably for the production of energy. It is now crucial to utilize natural resources as efficiently as possible without harming the environment. Applying the concepts of efficiency, dependability, continuity, and minimizing adverse environmental effects, energy policies of nations must be able to promote economic and social growth based on energy planning. Domestic resources should be used as quickly as feasible within the context of energy policy by adhering to these guidelines. These are metrics used to gauge each nation's degree of development. In this situation, sustainable energy policies need to be implemented. Sustainable energy is a concept that calls for the high efficiency generation of energy from all main energy sources. The comparison of fossil fuels with ecologically favorable via new technologies, the utilization of waste energy as input in another cycle, and the integration with economic development are made possible thanks to this clean technology. In conclusion, CCT is crucial for preserving the ecological balance of the planet and safeguarding the security of nations' energy supplies. The use of clean coal technology aims to lessen the negative environmental effects of coal use. The adoption of these technologies reduces coal combustion emissions while increasing the amount of energy that can be extracted from each tonne of coal used.

2.3 Hydrodesulfurization

In refineries, a process known as hydrodesulfurization is used to lower the sulfur level of distilled or fossil fuels (15 ppm). Since organic sulfur can only be extracted from organic molecules when the chemical connections are broken, removing organic sulfur from fossil fuels is challenging. But for this chemical rupture to occur, tremendous temperatures and pressures are required, and the circumstances are intimately related to the kind of hydrocarbon. HDS may happen at 200–425 °C and 150–250 pressure, depending on the hydrocarbon and the effectiveness of the desulfurization procedure used. Even though HDS technology can remove a significant quantity of simple inorganic and organic sulfur from oil, thiophenes like dibenzothiophene and its alkylated counterparts are especially resistant to HDS. However, there are significant drawbacks to the HDS technology that restrict its use. Among them, we can point out that H_2S produced by physical-chemical processes is harmful and shortens the equipment's useful life. Another drawback is the need for a significant capital investment as a result of a higher operating cost. Moreover, the value of the fuel produced by the treated oil is decreased when crude oil fractions are exposed to very high temperatures and pressures (over 360 °C). Last but not the least, the HDS atmospheric hydrogen causes a hydrogenation reaction of olefinic compounds, which lowers the fuel's calorific value. These compounds treated by HDS must travel through a new catalytic fluid to compensate for the loss of calorific value, which raises the cost. Given the drawbacks of HDS, BDS technology, similar to DBT, which has great potential for desulfurizing refractory compounds, is less expensive, does not harm the environment, and offers more benefits.

2.4 Oxidative Desulfurization

Oxidative desulfurization of coal is a process used to remove sulfur compounds from coal, which is important for reducing air pollution and complying with environmental regulations. Sulfur compounds in coal can contribute to the formation of sulfur dioxide (SO2) and other harmful pollutants when coal is burned for energy production. Oxidative desulfurization is an effective method for reducing the sulfur content of coal, but it may not remove all sulfur compounds. Some organic sulfur compounds may be more challenging to remove using this method, and achieving very low sulfur levels may require additional treatment processes or multiple passes through the desulfurization system. It's important to note that while oxidative desulfurization can significantly reduce sulfur emissions during coal combustion, it is not a complete solution to the environmental challenges associated with coal use. To further mitigate environmental impacts, technologies such as flue gas desulfurization (FGD) can be employed to capture sulfur dioxide emissions from power plants and industrial facilities. Additionally, transitioning to cleaner energy sources and improving coal combustion efficiency can help reduce the overall environmental footprint of coal-based energy production.

2.5 Study of the Application of ILs in Fuel Desulfurization

ILs are anion-and cation-based substances that are liquid at temperatures below 100 C. A wide range of liquid properties, low volatility, non-flammability, high thermal stability, and high solubility are just a few of the benefits that the majority of ILs have. Figure 2.3 presents examples of representative cations and anions. ILs have recently been used in coal chemistry to dissolve and swell coal and extract valuable compounds from the byproducts of coal liquefaction. The ability of non-polar organic solvents to dissolve or swell coal was, however, constrained because carboxylic groups may form ionic complexes or clusters in coal, resulting in high crosslinking density. According to studies, ILs can effectively swell, dissolve, and destroy coal structures as well as have some effect on coal's chemical reactivity. Several researchers had used ILs for coal desulfurization, especially for the removal of organic sulfur, in light of the impact of ILs on coal structure.

2.5.1 Analysis of the Impact of ILs on the Reactivity and Structure of Coal

Previous studies have shown that ILs dissolve and extract organic structures from coal by breaking not only the network structure but also hydrogen bonds and weak covalent bonds. Additionally, earlier findings showed that the expansion of coal can break weak hydrogen bonds and small molecules to reduce the free energy of coal structures, which increases the volume of coal samples and causes restructuring of the coal structures. Tetrahydrofuran (THF) was used as a swelling solvent by Rincon and Cruz to pretreat coal samples of various ranks. They discovered that low rank coal had a higher swelling rate and that after the coal was liquefied with tetralin, THF soluble matter and other hydrogen donor solvents performed better. When tetralin was used to swell coal at 100e250C under 1 MPa, it was observed that the solvent significantly improved coal's pyrolysis performance and could substantially increase the yields of volatile matter and tar. Additionally, [Emim][BF4] (1-ethyl-3-methylimidazolium tetrafluoroborate) was used to study the swelling of two coal samples, and it was found that the solvent's swelling capacity for coal samples peaked at 2.55 at 50e65 C. In order to prepare coal for swelling. The results revealed that the swelling pretreatment broke weak coal covalent bonds, improved coal swelling ratio with longer swelling times, and could recycle [Bmim]BF4 characteristics of five ionic liquids on four different types of Indian coal samples, due to the higher charge density with which [Bmpyr][CH3CO3] interacts with the polar portion of coal, resulting in an up to 1.5 [Bmpyr][CH3CO3] expansion ratio (as shown in Table 3). In general, the dissolving rate of IL to lignite was higher than that of bituminous coal because the concentration of oxygen-containing functional groups in lignite was higher than that in bituminous coal. Additionally, research has shown that DIMCARB (N,N-diaminammonium N0,N0 -dimethylcarbamate) is more soluble in triterpenoids and has high solubility for aliphatic groups in coal. Lignite had the greatest extraction rates and showed a linear connection with the carboxyl functional groups of the coal when various coals were dissolved in [Bmim]Cl (1-butyl-3-methylimidazolium chloride). Additionally, a different research hypothesized that [Bmim][H2PO4](1-butyl-3-methylimidazolium phosphate) interacted with oxygen-containing functional groups in Shengli lignite to enhance the carbonyl and ether bond groups of coal and alter the coal's pyrolysis performance. Bituminous coal samples were dissolved using six different imidazolium-based ionic liquids ([Hoemim][BF4], [Hoemim][BF4], [Amim][Cl], [Emim][AC], [Bmim][AC], and [Bmim][OTF]). It was discovered that the linked hydroxyl groups in coal were broken down into free hydroxyl groups and generated carboxyl groups as a result of the interaction between ILs and coal. The inert effect of the ionic liquid [Amim]Cl (chlorinated allyl3-methylimidazole) on low-temperature oxidation process was investigated. It was discovered that after treatment of coal, the content of hydroxyl groups, aliphatic methyl groups, methylene groups, and ether oxygen bonds related to hydrogen bonds decreased, indicating that IL had a dissolving and destructive effect on coal. The studies mentioned above showed that ILs are frequently used to dissolve and swell coal samples, showing that they can interrupt weak covalent bonds (ether bonds and aromatic carbon-carbon single bonds) as well as van der Waals and intermolecular forces in coal. As a result, the amount of hydroxyl functional groups in coal decreases, the oxygen-containing functional groups disintegrate, and the amount of fixed carbon increases. According to studies, ILs significantly altered the microstructure of coal. Additionally, studies on the direct liquefaction of coal residue to remove asphalt matter showed that ILs may rearrange the coal's microstructure in a way that affects the desulfurization process. Six different ionic liquid (IL) types were used: [Mmim] dimethyl 1,3-dimethylimidazolylphosphate (DMP), [Emim] 1-ethyl-3-methylimidazolylphosphate (DEP), [Bmim] dibutyl-3-methylimidazolium phosphate, [M3mpy] [DMP] (monophosphate of 1,3-dimethylpyridine), and [E3mpy]. To remove the asphaltenes from the direct coal liquefaction residue, [DEP] (1-ethyl-3-methylpyridinephosphate) and [MTEtA][DMP] (triethylmethylamine phosphate dimethyl ester) were used (DCLR). Table 4 shows that the desulfurization rate reached 96.94%, and the highest extraction rate of [MTEtA]DMP was up to 87.62%. To recover the asphalt material from DCLR, three alkyl sulfated ionic liquids (ILs) were used: [Emim][ESO4] (1-ethyl-3-methylimidazolium ethylsulfate), [Empy][ESO4] (1-ethyl-3-methylpyridinium ethylsulfate), and [E4N][ESO4]. The findings showed that the IL or the mixed reagent of an IL with an organic solvent may react with hydrogen bonds, p-p bonds, and other bonds in coal. The extraction yield of [E4N][ESO4]/NMP 14 1:2 achieved 46.2%. According to the literature mentioned above, ILs had a perfect impact on dissolving, swelling, and extracting organic compounds from coal. The

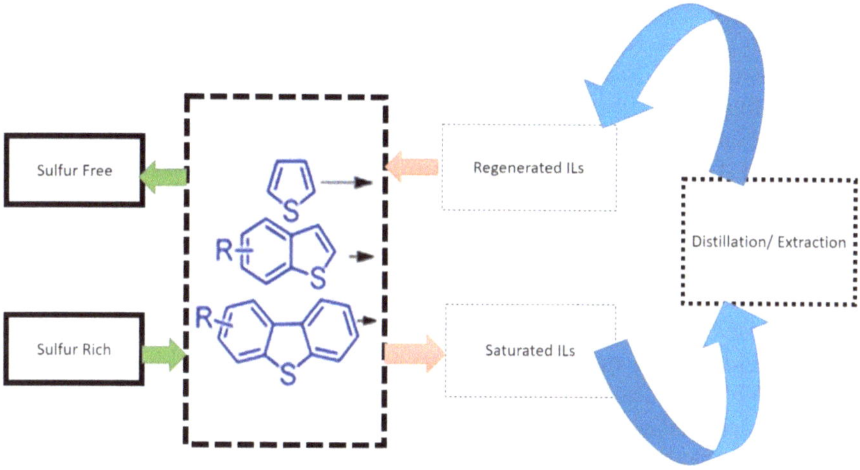

Fig 2.1 Ionic liquids for desulfurization [21]

findings of the experiment demonstrated that ILs may break down weak covalent connections in coal, such as hydrogen bonds, and have an impact on swelling during dissolution. Additionally, the same ILs had distinct effects on coal of various ranks. The sulfur level of coal following IL treatment was decreased to varied degrees, according to industrial and elemental tests. As a result, the ILs may be used to desulfurize coal and increase coal usage (Fig. 2.1).

2.6 Biodesulfurization of Coal

The structure of coal is intricate. One of the main causes of the poor rates of sulfur removal from coal has been the difficulty in accessing the various forms of sulfur in coal [33]. A combination of different inorganic and organic chemicals make up coal. The organic components are primarily made up of thiols, sulfides, disulfides, thiophenes, dibenzothiophene, and alkylated derivatives of dibenzothiophenes, while the inorganic portion is primarily made up of elemental sulfur, pyrites, H2S, and sulfate sulfur [141]. In order to achieve greater desulphurization rates, current research in the field of biodesulfurization is focused on the study of commercially accessible enzymes and highly effective microbial strains. The removal of organic sulfur from coal has trailed behind the industrial scale pyritic sulfur removal from coal utilizing microorganisms in studies throughout the years. Microbes with the ability to desulfurize DBT or similar chemicals should be researched as coal-desulfurization agents on an industrial scale since DBT makes up a large component of the organic sulfur. Furthermore, these researches may provide fresh data on a variety of possible microorganisms for efficient sulfur removal. A summary of the mechanism by which bacteria work to remove pyritic and organic sulfur from coal is shown in Fig. 2.4. Recently, considerable research has focused on using artificial neural networks (ANNs) to forecast the best process variables to achieve the highest coal desulfurization. In order to remove organic sulfur from coal taken from the Tabas coal deposit in Iran, *Rhodococcus* sp. was used, along with a mixed culture made up of *A. ferrooxidans* and *Leptospirillum ferrooxidans*. The findings of this study agreed with theoretical predictions of the removal of both organic and inorganic sulfur from coal [142]. Numerous acidophilic microorganisms, particularly the inorganic (pyritic) portion, have been reported to have the ability to desulfurize coal, in addition to the use of several heterotrophs (discussed in Sect. 2.2) [91]. The most extensively researched and strategically significant microorganism among the reported acidophiles is the proteobacterium *A. ferrooxidans* [143–146]. Table 4 provides a list of several acidophiles and their use in biodesulfurization. Pyrite dissolving from coal is an incredibly common occurrence. Numerous researches have been conducted employing a combination of *Acidithiobacillus thiooxidans* and *A. ferrooxidans* to dissolve coal's pyrite [148]. The amount of oxygen present, pH, the kind of iron sulfide present, the iron sulfide's reactive surface availability, and the microbial strain(s) all have a role in how quickly pyrite dissolves [154,155]. Recently, lignite and pit coal from the Halanga, Mintia, and Petrila mines were biodesulfurized using *A. ferrooxidans*. The effectiveness of coal biodesulfurization increased from 68.10 to 88.46% when the cultures were adapted to higher levels of ferrous sulfate (18 g/L). By raising the solid/liquid ratio from 5 g per L to 10 g per L, increased biodesulfurization efficiency was also accomplished. The improvement was from 57.30 to 76.41% for pit coal and from 72.21 to 82.50% for lignite [148]. To examine the impact of the configuration of organic matter in coal on biodesulfurization, the thermophilic acidophile *Sulfolobus solfataricus* ATCC 35091 was used in combination with the fungus strain *Penicillium chrysosporium* ME446 to desulfurize two low rank coals from Bulgaria and one from Turkey. Tests revealed that mature coals' organic component was resistant to microbial assault. However, *P. chrysosporium* in particular more rapidly attacked the organic material in lignite [151]. In a recent comparison study on the biodesulfurization and biodepyritization of coal, *A. ferrooxidans* was utilized. Total sulfur was eliminated in 16 days at a rate of 50.6%, of which pyritic sulfur made up 69.9%. Several analytical studies suggested that the microbial strain initially used massive pyrite preferentially. The microbially treated coal sample included jarosite [149]. Recently, it was discovered that a different species of the *Acidithiobacillus* genus, *Acidithiobacillus ferrivorans*, isolated from the Balya acid mine drainage, was effective in desulfurizing coal. In 14 days, a 33% sulfur reduction from the 500 + 250 m size fraction was accomplished. The surface of the coal particles had been altered, according to SEM and FTIR studies [152]. The Cayirhan lignite was studied in batch reactors with *A. ferrooxidans* and *A. thiooxidans*, and 78.2% of the sulfur was removed [147]. Cismasiu et al. [156] have recently researched the biochemical properties of the acidophilic microbial strain *Acidiphilium* that is engaged in the breakdown of both organic and inorganic substances. Based on their capacity to hydrolyze starch in the presence of different inorganic heavy metal ions, acidophilic strains that are responsible for the breakdown of organic compounds were chosen. After 21 days of incubation, it was shown that the acidophile's extracellular hydrolytic activity increased (by almost 50%–60%) in the presence of 2–3 g/L starch. This increase was also found when sulfate was present in the surrounding medium. As a result, this discovery employing the heterotrophic acidophile from the genus *Acidiphilium* recommended starch as a source of nutrients in the process of ecological restoration of heavy metal damaged places. Additionally useful in bioflotation are acidophiles. Amini et al.'s work [150] concentrated on the application of *A. ferrooxidans* in the bioflotation process to remove pyritic sulfur from coal. A first-order flotation model was used to compare the effectiveness of the microbe with that of sodium cyanide as a flotation agent. In the flotation of sulfide minerals, sodium cyanide is a widely used depressant for pyrite. *A. ferrooxidans* is a helpful agent in the biodepyritization of coal due to its capacity to inhibit the flotation of pyrite. Additionally, it was discovered that the *A. ferrooxidans* LY01 cells were more effective in removing sulfur from the environment when they were adapted to pyrite rather than ferrous iron. In comparison to cells that had been acclimated to ferrous iron, pyrite-adapted cells produced five times more EPS. This demonstrates the value of substrate specificity, which is important for biodesulfurization [153].

Kodama and associates were the first to describe the oxidative cleavage of DBT. They postulated the well-known Kodama route and discovered a number of intermediates formed during the oxidation. This route involves the three key processes of hydroxylation, ring cleavage, and hydrolysis (Fig. 1.4). Many different microbes have been found to use this route. The *Pseudomonas jijani* and *P. abikonesis* bacteria, which serve as the cornerstone of the Kodama route, were two that were described in 1970. The breakdown of DBT in these microbes is encoded by a 55-MDa plasmid, as shown by Monticello and colleagues in 1985. Both cell development and more DBT oxidation are inhibited by the byproduct of DBT oxidation. According to studies, succinate inhibits DBT oxidation, whereas both naphthalene and salicylic acid speed up the process. DBT itself has a considerably smaller impact on the process. *P. putida* also uses a different mechanism, in which DBT is converted to DBT sulfone, in addition to the Kodama process. The DBT is initially metabolized by the organism via the Kodama route, and then it is transformed by the DBT sulfone pathway (Fig. 1.5). The fungus *Cunninghamella elegans* and *Pleurotus ostreatus*, which oxidized DBT to DBT sulfoxide and DBT sulfone as dead end products, were also found with similar discoveries. The similar mechanism has also been linked to *Beijerinckia* sp. and *Rhizobium meliloti*. In the case of *Beijerinckia* sp., pyruvate and 3-hydroxy-2-formyl benzothiophene are also produced (HFBT). There have been occasions when alternative potential HFBT products have been identified, despite the fact that HFBT is often thought of as the byproduct of DBT breakdown through the Kodama route. In 1994, Eaton and Nitterauer demonstrated that benzothiophene-2,3-dione was created from 2-mercaoptophenylglyoxylate by acid-catalyzed dehydration since acidic conditions were common when culture supernatants were extracted. Furthermore, it has been observed that HFBT may

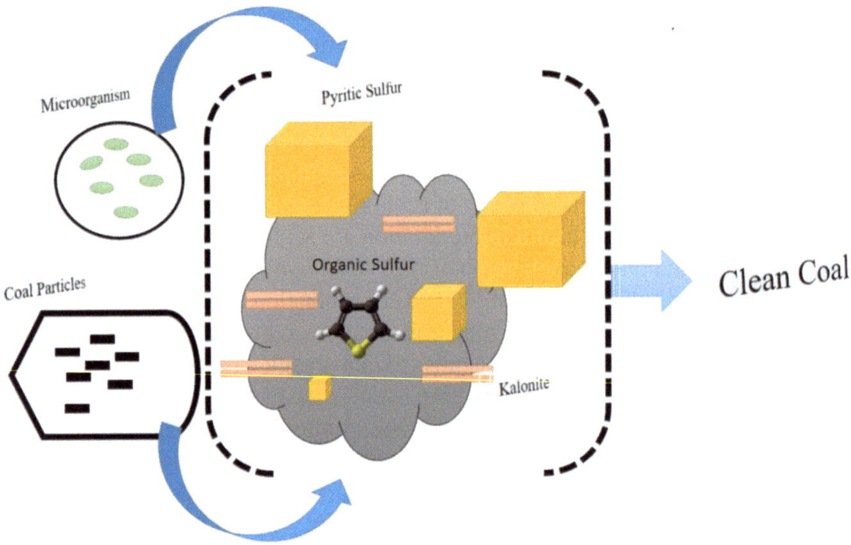

Fig. 2.2 Bio desulfurization of coal [22]

break down into carbon dioxide. The replaced DBTs are more prevalent in HDS-treated diesel. Due to this, the aerobic metabolism of methyl DBTs in *Pseudomonas* strains was also investigated. These strains converted DBT into benzothiophene-2,3-dione, HFBT, DBT sulfoxide, and sulfone as byproducts. It was discovered that when methyl DBT underwent Kodama desulfurization, the carbon-carbon cleavage happened on the benzene ring that lacked a substitution group, resulting in the formation of the matching methyl 3-hydroxy-2-formyl-benzothiophene (methyl HFBT). To explore the transformation of dimethyl DBTs, pure cultures of the *Pseudomonas* strains BT1, W1, and F and four petroleum-degrading mixed cultures were employed. The unsubstituted rings were broken down in 3,4-dimethyl DBT to produce several compounds, including 6,7-dimethyl HFBT and 6,7-dimethyl DBT-2,3-dione. When any of the cultures employed 4,6-dimethyl DBT as substrate, no discernible C-C cleavage was seen. This indicated that, in contrast to their methylation derivates, the Kodama pathway enzymes preferred to cleave the unsubstituted homocyclic rings. Research on the HFBT, DBT sulfoxide, and DBT sulfone degradation by mixed culture revealed a decrease in HFBT. In the case of DBT sulfoxide and DBT sulfone, no sulfate was released into the medium, but carbon dioxide was produced instead, indicating that these compounds had degraded but had not completely demineralized (Fig. 2.2).

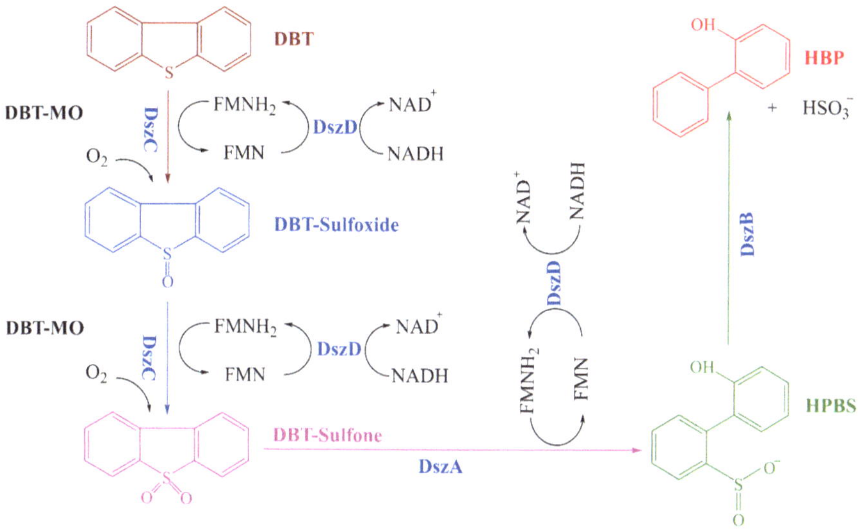

Fig 2.3 C-S cleavage of organic sulfur into 2-HBP [24]

2.6.1 C–S Cleavage (Anaerobic Biodesulfurization)

A reducing equivalent is needed for reductive desulfurization, which converts DBT to biphenyl and releases the sulfur as hydrogen sulfide. Dibenzylsulfide, an organosulfur molecule, is anaerobically desulfurized by *Desulfovibrio* sp. to toluene, benzyl mercaptan, and hydrogen sulfide in the presence of molecular hydrogen. Any other source may likewise supply the decreasing equivalent. A bioelectrochemical procedure was created to supply electrons to D. desulfuricans M6 using electrochemical cells, resulting in the production of biphenyl and hydrogen sulfide from DBT. Numerous studies have also looked at how hydrocarbons affect the decomposition of reductive DBT. Biphenyl and hydrogen sulfide were the byproducts of desulfurization when DBT was dissolved in 100% dimethylformamide, an organic solvent. The desulfurization of organosulfur compounds was one factor in the creation of hydrogen sulfide, but it was not the only one. *Desulfomicrobium scambium* and *Desulfovibrio longreachii* are two examples of anaerobic bacteria that have been found to only be able to desulfurize 10% of DBT dissolved in kerosene. Five unidentified products were seen, and the mechanism used to produce them was reportedly distinct from the one used to produce the final product biphenyl. In anaerobic environments, there is less oxidation of hydrocarbons to undesirable chemicals including colored and gum-forming products. This benefit might serve as motivation to carry out further study on reductive biodesulfurization. However, sustaining an anaerobic process is quite challenging, and most isolated strains' specific activity has been observed to be minimal for DBTs (Fig. 2.3).

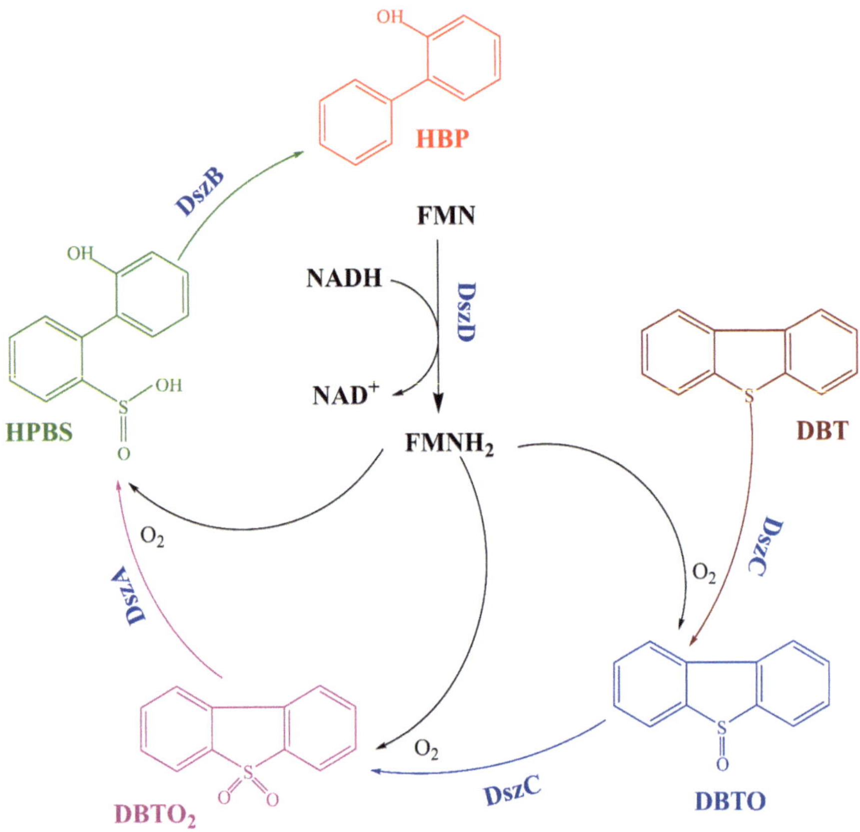

Fig 2.4 Oxidative desulfurization of sulfur compounds via C-S cleavage [25]

2.6.2 *Oxidative C-S Cleavage (Specific Oxidative Desulfurization)*

The calorific value of the fuel is kept because in a specialized oxidative desulfurization process sulfur from DBT and methyl DBT is removed in a sulfur-specific way without damaging the carbon skeleton. Due to its selectivity for sulfur atoms and ability to operate under aerobic circumstances, studies on biodesulfurization have concentrated on this method over the last 10 years. The initial use of this sulfur reduction method was for *Rhodococcus*. Several additional aerobic bacteria, including *Rhodococcus erythropolis* and *R. rhodochrous* IGTS8, are also present. *Gordona* sp., *Sphingomonas* sp., *Corynebacterium* sp., *Mycobacterium* sp., *Bacillus subtilis*, *Nocardia* sp., and *Paenibacillus* sp. have been reported to be able to desulfurize DBT via a sulfur-specific pathway (Fig. 2.4).

2.6.3 Oxidative C-S Cleavage

The calorific value of the fuel is kept because in a specialized oxidative desulfurization process sulfur from DBT and methyl DBT is removed in a sulfur-specific way without damaging the carbon skeleton (Fig. 2.4). Due to its selectivity for sulfur atoms and ability to operate under aerobic circumstances, studies on biodesulfurization have concentrated on this method over the last 10 years. Gallagher and coworkers originally described this sulfur removal method for *Rhodococcus rhodochrous* IGTS8 in 1993. Four distinct sulfur-containing compounds are generated during the conversion of DBT to 2-hydroxybiphenyl (2-HBP), which is why this route is referred to as the 4S pathway. As illustrated in Fig. 1.6, this mechanism involves the sequential oxidation of DBT sulfur to produce the products sulfoxide (DBTO), sulfone (DBTO2), sulfinate (HPBS), and hydroxybiphenyl (HBP). Although various genera have been identified as having distinct desulfurization activity to yet, the majority of them are *Rhodococcus* genera. Other aerobic bacteria include *R. rhodochrous* IGTS8, *Gordona* sp., *Sphingomonas* sp., *Corynebacterium* sp., and *Mycobacterium* sp.

2.7 Recent Developments in Sulfur Removal Techniques for Coal

The primary barriers to the employment of biological processes on an industrial scale, as previously indicated, are that they require a long time compared to other processes and may be used at low solid ratios. Another issue is that these creatures could have difficulties adapting to their surroundings as a result of the challenges that arise from the nearby living things. By separating organisms from their surroundings, adaption may be prevented. Microorganisms are capable of adapting to shifting environmental factors. According to reports in the literature, native microorganisms isolated from coal may be helpful for coal biodesulfurization. Significant desulfurization rates may be attained in low-cost settings employing isolated microbes from coal itself. If desulfurizing microorganisms are present in coal naturally, isolating and using them to produce clean coal seems like a viable strategy since it avoids issues like adaptation. To find microorganisms appropriate for biodesulfurization procedures, Gomez and colleagues studied the microbial ecology of three different Spanish coals. They discovered promising findings, with the B12 strain isolated from coal sample B removing up to 67% of sulfur. However, both microbial diversity and coal heterogeneity might lead to serious operational issues. The endogenous fora contained within the feed coal in a biodesulfurization plant might potentially produce undesirable interactions in the working culture since it would not be practicable to sterilize the feed coal in a biodesulfurization plant. Although biological methods have drawbacks related to time and solid rate, they offer benefits over many other processes in that they are cost-effective and ecologically benign. It is anticipated that research is being done to solve these issues. The studies conducted to address these issues may be categorized under a number of different headings:

1. Providing a detailed description of the desulfurization methods of microorganisms and outlining their metabolism to improve efficiency. Numerous microbes have the ability to desulfurize. Additional factors that have helped to realize biodesulfurization at large-scale applications include the improvement of knowledge of microbial ecology, metabolic engineering, comprehension of new biochemical pathways and metabolic reactions, improvement of instrumental analyses required for characterization, and development of reactor systems.
2. Using microorganisms that have been genetically altered. Genetically modified microbes to remove both inorganic and organic sulfur forms from coal are now the subject of intensive research due to their ability to minimize treatment times and provide more consistent outcomes. The regulation of the intricate processes brought on by an increase in microbe density is this approach's biggest drawback. Isbister and Doyle described a mutant *Pseudomonas* sp. CB1 pure culture and patented a method for using this culture to extract organic sulfur compounds from carbonaceous substrates like coal and petroleum. Initial findings revealed that attempts to grow the microorganisms together failed and that sulfur removal using mixed cultures was not repeatable. However, sulfur removal was somewhat effective when CB1 and CB2 strains were employed sequentially. *Acidithiobacillus* that has been genetically modified may be used for coal bioleaching, according to Koizumi. In order to create a mutant strain of *A. ferrooxidans*, multi-stage mutagenesis agents like EMS and UV radiation were used. Recombinant *Pseudomonas putida* strain was used to desulfurize dibenzothiophene. Desulfurization genes from *R. erythropolis*, such as dsz A, dsz C, and dsz B, were cloned in the recombinant strain of *Pseudomonas* by creating a plasmid, which improved dibenzothiophene desulfurization. Martinez and colleagues discovered that adding co-substrates like acetic, citric, and succinic acids could improve the biodesulfurization of *Pseudomonas putida* CECT5279, a genetically engineered organism. However, using acetic acid at a concentration of 1.5% produced the best improvements in the biodesulfurization. Both the genes hpaC and dszABC from *E. coli* W and *R. erythropolis* IGTS8 were present in *Pseudomonas putida* CECT5279. They said that under ideal circumstances, the procedure's yield may rise to 140%. A synthetic gene producing a peptide (Sulpeptide 1) with a high percentage of methionine and cysteine residues was introduced into *Rhodococcus erythropolis* IGTS8 to serve as a sulfur sink, according to a research by Wang et al. Microorganisms' capacity for biodesulfurization increased as a consequence of this genetic modification. *Rhodococcus qingshengii* was engineered to express a peptide that serves as a sulfur sink

(Sulpeptide 1), and over the course of around 400 generations, adaptive evolution was effective in significantly improving both its metabolism of DBT. The adaptive evolution contributed more to these increases than the effect of Sulpeptide 1 expression.

3. Sequential application of several procedures. Recent years have seen a strengthening of the idea that employing many processes sequentially produces more effective outcomes than using each process alone. Sequential applications using gravity, flotation, chemical treatments, and biobeneficiation are viable options for removing sulfur and may make use of other factors affecting the low-rank coal's quality. Attia investigated the viability of selective biomodification of pyrite surface to improve pyrite frothability in coal desulfurization. In order to enhance the removal of pyrite from coal, bioconditioning using bacterial culture was used in this research before froth flotation. Nearly 95% of the pyritic sulfur was removed over a 6-week bioconditioning period prior to flotation, which included 30 hours of conditioning rain. In the investigation by Stevens et al., coal was depyritized at a larger laboratory scale using a combination of bacterial (*A. ferrooxidans* and *Leptospirillum ferrooxidans*) and physical approaches. The results show that at the conditions of a 5-day reactor residence time and a 20% slurry concentration, pyritic sulfur was reduced by 87%, while ash was reduced by 36%, with 90% recovery of heating value. In addition to removing over 30% of the inorganic minerals, a combination of bacterial and physical techniques removed over 90% of the pyritic sulfur. According to the mechanism for microbial flotation using *A. ferrooxidans* for pyrite suppression in a different study, microbial column flotation should be used for both ore dressing and desulfurization of fuel-neutral coal in order to achieve high purity. *Pseudoclavibacter* sp. can be used to demonstrate multi-step treatments for coal desulfurization with bioprocessing and flotation. These procedures include bioflotation and biomodification (also known as biooxidation) (column flotation). Extracellular polymeric substances produced by *Pseudoclavibacter* sp. strain SKC/ XLW-1 may be used to biodesulfurize organic sulfur from Tondongkura coals in Indonesia. Turkish lignite samples from the Mihaliccik Region were treated by Aksoy and colleagues using a combination of biodesulfurization, gravity, and flotation procedures. A multi-gravity separator (MGS) was used for gravity separation, and a Denver flotation cell was used for flotation. With the help of the fungus *Alternaria* sp. Cf1, the concentrates from the MGS and flotation tests were further biodesulfurized (accession number KF564051). After MGS treatment and biodesulfurization, the overall sulfur content dropped from 3.25 to 1.56%, whereas after flotation treatment and biodesulfurization, it dropped from 3.55 to 1.81%. The same fungus, *Alternaria* sp. Cf1, was recently used by Koca et al. to desulfurize the lignites in the Cayirhan area of Turkey. Additional biodesulfurization tests using the fungus *Alternaria* sp. Cf1 were carried out after physicochemical

cleaning by flotation to reduce ash content. For the concentrates from the flotation and microbiological treatments, the sulfur emission values decreased from 12.76 to 1.01 to 0.34 for the head sample, flotation, and microbial treatment, respectively. This represents a reduction of 97.34% on a dry ash-free basis. Displays pictures taken using a scanning electron microscope (SEM) of the surface of lignite before and after biodesulfurization. The SEM method has typically been employed in coal biocleaning investigations to indicate the structure of the coal surface and the morphology of the microorganisms that are attached to the coal, such as fungi or bacteria. Compared to other microscopy methods like light microscopy and confocal laser microscopy, it may be used to examine early cell attachment, colonization, or the development of biofilm at a high resolution and magnification. The photos show that the coal pyrite surface was smoother before treatment, but that the pyrite surface was affected by flotation. Additionally, after 12 days of microbial treatment, considerably rougher, nearly no smooth surfaces were formed because the cells were well adherent. In addition to SEM, transmission electron microscopic observations revealed a relationship between coal particles and bacterial cells. Coal dust particles came into touch with certain microbial cells, adhering to their cell walls. The presence of fibrous and highly hydrated layers around the cell wall suggested that the majority of cells produced bacterial extracellular polymeric material (EPS). Organic materials with a greater water content make up EPS. The EPS is hydrated, and holes surrounding the bacterial cells are then visible. It is also expected that the EPS produced by microbes may boost interaction between bacteria and hydrophobic water-insoluble substances like coals and increase the surface area of coals and their bioavailability, enhancing coal biodesulfurization. This explanation exhibits parallelism with SEM images of coal particle surfaces, demonstrating the formation of fine pores on coal surfaces as a result of bacterial action, with the EPS (metabolites) of the bacteria possibly mediating the microbial attachment to coal surfaces, allowing the bacteria to oxidize and desulfurize the sulfur in coal. For a few decades, biological processes have been thoroughly investigated on a laboratory scale in conjunction with other cleaning techniques. For instance, biological procedures have been applied before physical or physicochemical processes (such as flotation) to boost the efficacy of coal flotation or after these processes to increase the efficacy of desulfurization. These researches claimed that adding a biodesulfurization step boosted the efficiency of sequential processes. Figure 2.3 depicts graphically how biological processes work in conjunction with other clean coal technologies. The qualities of the coal itself determine the stage at which biological activities are to be introduced into the system. Iron oxide (hematite), iron hydroxide, and silica are additional components of coal that cause environmental issues in addition to sulfur. The silicon impurities in particular raise the cost of coal transportation. In future strategies, improving coal purity may also help to increase the value of coal and reduce environmental pollution, in addition to the desulfurization process. Growing and dormant cells of heterotrophic *Rhodococcus erythropolis* strains PD1, R1, and

FMF and *R. qingshengii* were used in the research by Etemadifar et al. for sulfur removal and bioleaching of iron and silica from coal. *R. erythropolis* strain PD1 is an excellent option for additional research on coal bioleaching in this study since it not only decomposes organic sulfur. *Exophiala spinifera* FM and *Fusarium oxysporum* FE, two previously researched fungi, have the capacity to extract sulfur, iron, and silicon from coal in both growing and resting cells. In addition to these methods, which include employing genetically modified organisms and sequential treatment, other methods such as altering the medium or using biphasic media were also investigated. For instance, biphasic media design has been used to improve upon single-phasic media's poor conversion rate. Some biodesulfurization experiments' poor conversion rates may in part be explained by the mass transfer effect, which prevents some microbes from evolving for an extended period of time in mono-phasic aqueous medium emulsions. In particular, it has been assessed how sensitive *P. putida* CECT5279 is to the DBT mass transfer constraint in this environment. According to some research, using supplements and additives as catalysts may increase biodesulfurization by facilitating mass transfer between the organic and aqueous phases. Sui and associates used a nanoenzyme to coal using a novel biochemical method. The enzyme treatment altered the structure of coal. Due to the breaking of C-C/C-S/C-N bonds caused by enzyme treatment, the H/C ratio rose. Compared to emissions from the raw coal tests, burning the enzyme-treated coal reduced NOx emissions by 16.72–21.43% throughout the course of the whole test period and SO_2 emissions by 5.42–7.31% when operating at 85% load.

2.8 Conclusion and Future Recommendation

In conclusion, the techniques for sulfur removal from coal play a pivotal role in addressing the environmental challenges associated with coal combustion. The high sulfur content in coal contributes to the release of sulfur dioxide (SO2), a major precursor of acid rain and air pollution. Therefore, implementing effective sulfur removal methods is crucial for mitigating these negative impacts. Several techniques have been developed to achieve this goal, each with its own set of advantages and limitations. Physical methods such as gravity separation, flotation, and magnetic separation have been employed to remove sulfur-bearing minerals from coal. While these methods can effectively reduce sulfur content, they often require complex equipment and are not always selective in targeting specific sulfur compounds. Chemical techniques involve using reagents to react with sulfur compounds and convert them into soluble forms that can be separated from coal. Processes like coal washing and chemical leaching have been utilized. However, they may generate waste products and require careful handling of chemicals. Biodesulfurization, or microbial desulfurization, utilizes microorganisms to metabolize sulfur compounds in coal. This environmentally friendly approach operates under milder conditions and can selectively target specific sulfur compounds, making it a promising option for coal desulfurization. Hydrothermal techniques involve treating coal with water and high temperatures under controlled conditions. These processes can effectively remove both organic and inorganic sulfur, but they require energy-intensive procedures and can affect coal properties. Pyrolysis and gasification involve the thermal degradation of coal under controlled environments. These processes can reduce sulfur content, but they often produce additional byproducts that need further treatment. Oxidative methods use oxidizing agents to convert sulfur compounds into soluble forms. These methods can be effective for specific sulfur compounds, but they might also involve complex reaction pathways. Advanced combustion technologies and gas cleaning processes can capture sulfur compounds from flue gases after coal combustion. These methods are crucial for reducing emissions but do not directly affect the sulfur content in coal.

In practice, a combination of these techniques might be necessary to achieve the desired level of sulfur removal while considering the specific coal composition, economics, and environmental concerns. The choice of technique depends on factors such as the type of sulfur compounds present, the desired level of desulfurization, process efficiency, and environmental considerations. Ultimately, the ongoing research and development of sulfur removal techniques hold promise for enhancing the sustainability of coal utilization. As the energy landscape evolves toward cleaner alternatives, advancements in sulfur removal technologies will contribute to reducing the environmental footprint of coal-based energy generation.

References

1. Abas N, Kalair A, Khan N (2015) Review of fossil fuels and future energy technologies. Futures 69:31–49
2. Mohajan H (2018) Acid rain is a local environment pollution but global concern. Open Sci J Anal Chem 3:47
3. Boudou J, Boulegue J, Malechaux L, Nip M, De Leeuw J, Boon J (1987) Identification of some Sulphur species in a high organic Sulphur coal. Fuel 66:1558–1569
4. Dai S, Hower JC, Finkelman RB, Graham IT, French D, Ward CR et al (2020) Organic associations of non-mineral elements in coal: a review. Int J Coal Geol 218:103347
5. Dai S, Finkelman RB, French D, Hower JC, Graham IT, Zhao F (2021) Modes of occurrence of elements in coal: a critical evaluation. Earth Sci Rev 222:103815
6. Çelik PA, Aksoy DÖ, Koca S, Koca H, Çabuk A (2019) The approach of biodesulfurization for clean coal technologies: a review. Int J Environ Sci Technol 16:2115–2132
7. Zhang M, Hu T, Ren G, Zhu Z, Yang Y (2017) Research on the effect of surfactants on the biodesulfurization of coal. Energy Fuel 31:8116–8119
8. Hu T, Yang Y, Zhang M, Gao Y, Cheng Q, Ji H (2019) Biodesulfurization of coal using Rhodococcus erythropolis SX-12 and Acidithiobacillus ferrooxidans GF: a two-step approach. Energy Sci Eng 7:162–169
9. Yang X, Wang S, Liu Y, Zhang Y (2015) Identification and characterization of Acidithiobacillus ferrooxidans YY2 and its application in the biodesulfurization of coal. Can J Microbiol 61:65–71

10. Ishfaq HA, Banerjee A, Qamar S (2021) Bio-desulfurization of coal using biotechnological approach, making coal a less harmful fuel. In: Clean coal technologies: beneficiation, utilization, transport phenomena and prospective. Springer, pp 161–179
11. Liu T, Hou JH, Peng YL (2017) Biodesulfurization from the high sulfur coal with a newly isolated native bacterium, aspergillus sp. DP06. Environ Prog Sustain Energy 36:595–599
12. Koyunoğlu C, Karaca H (2023) Microbial desulphurisation of coal: a review. Int J Sustain Energy 42:1–24
13. Shah SA, Asadullah ZN, Qaisrani I, Khan A, Naeem AN, Mengal et al (2024) Microbial desulphurization of low-ranked coal using Pseudomonas aeruginosa. Chem Africa 7:1–10

Chapter 3
Microorganisms for Sulfur Removal from Coal

Abstract It is very important to remove sulfur from coal before combustion in order to prevent the release of sulfur oxides (SOx). The inorganic sulfur component of coal has been removed using both chemical and physical techniques. This element is typically connected to pyrites (FeS2). However, no commercially viable techniques for the removal of the organic sulfur component have been identified. This article offers microbiological techniques for releasing sulfur bound to organic matter from coal. This study's specific objectives were to determine whether a strong biologically active population, such as activated sludge biomass, can remove organic sulfur from coal and to develop sulfur analytical techniques that are less time-consuming than the established norms for screening research results.

3.1 Introduction

One of the most widely used, affordable, dependable, heterogeneous, noncrystalline, and naturally abundant nonrenewable fossil fuels is coal [1]. Fuel scientists were forced to develop coal beneficiation strategies for effective sulfur reduction due to the gradual loss of high-quality coal and the release of significant amounts of SO_2 into the atmosphere during industrial burning, which led to atmospheric pollution, acid rain, and eventually harmful effects on the environment [2–4]. Anthracite, bituminous, subbituminous, and lignite coal are the different types of coal. The features of coal, such as the amount of carbon, volatile matter, and heat content, greatly influence how coal is classified as "high rank" or "low rank." Coal contains sulfur, which can be either organic or inorganic (pyritic and sulfide in nature), and regulates the amount of energy produced [5]. The removal of organic sulfur from coal of low quality is thought to be exceedingly challenging. Different coal kinds' high sulfur content has restricted their industrial use. Since environmental conditions have a significant impact on the sulfur removal from coal based on live organisms, optimizing various parameters at the laboratory size is a necessary step before achieving clean coal at the industrial scale [6]. When attempting to reduce the amount of sulfur in various coal kinds, it is vital to take into account temperature,

A. S. Jatoi, *Biological Removal of Sulphur from Coal*, Green Energy and Technology, https://doi.org/10.1007/978-981-96-8628-5_3

pH, moisture content, nutritional media composition, coal types, particle size, pulp density, and the microorganisms used. In order to better explore the process at an industrial scale, the cost-benefit ratio of the biodesulfurization process in comparison to other physicochemical approaches would be useful. With an emphasis on biological desulfurization, various microbes that remove sulfur, the mechanism of biodesulfurization, the impact of various factors on sulfur elimination, and future perspectives, we have attempted to briefly discuss the various processes for sulfur removal from coal in this chapter. Applications and necessitates reducing sulfur content using appropriate techniques. Desulfurization, often known as the removal of sulfur from coal, can be carried out using a variety of physical, chemical, and biological techniques. Flotation, flocculation, gravity separation, sedimentation, ultrasonication, microwave irradiation, and pyrolysis are a few of the often employed physical techniques for desulfurization that are less expensive but have not been very successful in removing organic sulfur. In order to quickly remove sulfur from coal matrix, various acids and bases are applied, allowing for either oxidation or reduction. It has been extensively documented to treat waste with sodium borohydride, potassium hydroxide, hydrochloric acid, nitric acid, subcritical water, and supercritical liquids like methanol and ethanol. The physicochemical procedures are pricy, may result in secondary pollutants that worsen the environment, are challenging to apply, and are not environmentally friendly. Comparatively, the biological agent-based sulfur removal from coal (biodesulfurization) is cost-effective and environmentally safe, produces a negligible amount of SO_2 release into the atmosphere, and even has the potential to increase the coal's overall calorific value. It's interesting to note that the biodesulfurization process, which is a clean coal technology, has proven to be quite effective in removing sulfur from low-rank coal. Bacteria, actinomycetes, and fungi are some of the biological agents that have been successfully utilized to remove sulfur from coal from a variety of coal types. According to research on the model organic sulfur compound dibenzothiophene, the crucial stage in the removal of sulfur from coal entails the oxidation of the sulfur moiety and the breaking of both the C C and C S bonds (DBT). Sulfur removal procedures can be either aerobic or anaerobic in nature. Notably, autotrophic or heterotrophic microbes may aid in the desulfurization process. The monooxygenase, desulfinase, and FMN-oxidoreductase enzymes that contribute to desulfurization have been found and are expressed as "DszC and DszA," "DszB," and "DszE," respectively.

3.2 Current Scenario

One of the most widely used nonrenewable primary fossil fuels for energy generation is coal, which is also one of the least expensive, dependable, heterogeneous, noncrystalline, and naturally abundant fossil fuels. Fuel scientists have been forced to develop coal beneficiation strategies for effective sulfur reduction due to the gradual loss of high-quality coal and the release of significant amounts of SO_2 into the atmosphere during industrial burning, which ultimately results in atmospheric pollution, acid rain, and adverse effects on the environment. There are four types of coal: lignite, subbituminous, bituminous, and anthracite. The features of the coal, such as its carbon concentration, volatile matter content, and heat content, play a significant role in how coal is classified as "high rank" or "low rank." The amount of sulfur in coal, which can be either organic or inorganic (pyritic and sulfide in nature), affects how much energy is produced. It is thought to be particularly challenging to remove organic sulfur from low-quality coal. Different varieties of coal's high sulfur content have put a cap on its industrial use. The optimization of many parameters at the laboratory scale is a necessary step before producing clean coal at the industrial scale since the sulfur removal from coal based on living organisms is greatly influenced by environmental conditions. When attempting to reduce the amount of sulfur in different forms of coal, it is vital to take into account temperature, pH, moisture content, the makeup of the nutritional media, the types of coal, particle size, pulp density, and the microorganisms used. A deeper understanding of the process at an industrial scale would result from comparing the cost-benefit ratio of the biodesulfurization process to other physicochemical approaches. We have attempted to quickly examine the many methods for removing sulfur from coal in this chapter, with a focus on biological desulfurization, various bacteria that remove sulfur, the mechanism of biodesulfurization, the impact of various parameters on sulfur elimination, and future prospects. Applications and demand that sulfur content be reduced using appropriate techniques. Desulfurization, a word used to describe the process of removing sulfur from coal, can be accomplished through a variety of physical, chemical, and biological techniques. While less expensive, some of the regularly used physical desulfurization techniques, like flotation, flocculation, gravity separation, sedimentation, ultrasonication, microwave irradiation, and pyrolysis, have shown to be ineffective at removing organic sulfur. The chemical treatment uses a variety of acids and bases to quickly remove sulfur from coal matrix by either oxidation or reduction. The use of potassium hydroxide, sodium borohydride, hydrochloric acid, nitric acid, subcritical water, and supercritical liquids like methanol and ethanol for treatment is widely known. The physicochemical procedures are pricy, potentially environmentally degrading causing secondary pollutants, difficult to apply, and not ecologically sound. The biological agent-based sulfur removal from coal (biodesulfurization) is, in contrast, cost-effective and environmentally safe, resulting in a small release of SO_2 into the atmosphere, and even has the potential to increase the overall calorific value. It's interesting to note that low-rank coal's sulfur content can be effectively removed by the biodesulfurization process, a clean coal technology. In addition to other biological agents, bacteria, actinomycetes, and fungi have been reported to successfully remove sulfur from different types of coal. According to research on the model organic sulfur compound dibenzothiophene, oxidizing the sulfur moiety and rupturing the C and C S bonds is a crucial step in the process of removing sulfur from coal (DBT). Both aerobic and anaerobic processes can be used to remove sulfur. It's important to note that either autotrophic or heterotrophic microbes may aid in the desulfurization process. Monooxygenase, desulfinase, and FMN-oxidoreductase have been discovered as the enzymes involved in desulfurization and are expressed as "DszC and DszA," "DszB," and "DszE," respectively.

3.3 Role of Bacteria in the Removal of Sulfur from Coal

The fact that coal is highly hydrophobic and contains aromatic sulfur compounds renders it resistant to microbiological desulfurization. Nevertheless, numerous microorganisms connected to desulfurization are identified in nature [6]. These bacterial species may be used as an environmentally beneficial method to regulate the sulfur poisoning of the atmosphere brought on by the combustion of fossil fuels like coal. Effective methods for removing sulfur from coal include microorganisms with a high rate of reproduction, the capacity to endure challenging environmental conditions, and tolerance to higher concentrations of heavy metals contained in coal matrix. Microbiological processes used to desulfurize coal are wet processes. The concurrent microbiological desulfurization by iron oxidizers might be direct or indirect in nature. Pyrite oxidation by interacting microbes occurs during the direct action during inorganic sulfur rejection, but during the indirect action process, ferric iron produced by bacterial metabolism catalyzes the oxidation of pyrite. To date, it has been reported that bacteria from the genera *Bacillus*, *Acidithiobacillus*, *Mycobacterium*, *Sulfolobus*, *Corynebacterium*, *Leptospirillum*, *Arthrobacter*, *Sphingomonas*, *Burkholderia*, *Rhodococcus*, *Pseudomonas*, *Desulfovibrio*, *Serratia*, and *Stenotrophomonas* significantly contribute to the removal of inorganic and coal surface features alter as a result of bacterial action, but energy generation capacity is unaffected. Electron microscopy, infrared spectroscopy, X-ray diffraction, X-ray photoelectron spectroscopy, X-ray absorption near edge spectroscopy, and elemental analysis are excellent tools for observing these changes in coal matrix. Recently, a method for using *Acidithiobacillus ferrooxidans* to catalyze sulfur removal from coal was presented. The desulfurization ability of the chosen microorganism was demonstrated to be impacted by the desulfurization investigation carried out in three batches in medium M9K with and without iron. Pyritic and total sulfur were removed from several sets of iron-containing media at rates of 55.6 percent and 10.0%, 77.1% and 16.1%, and 86.5% and 28.2%, respectively. The stated bacterium's ability to desulfurize was enhanced by the addition of iron-free M9K media; nevertheless, a higher amount had a negative impact on the microbiological sulfur removal phenomena. Another explanation for the decrease in desulfurization was the presence of calcium compounds like $CaCO_3$ and CaO. Through the use of bacteria extracted from coal, an experimental study has demonstrated a good effect on the calorific value of biologically desulfurized coal, increasing it by 19%. After bacterial treatment, a total of 70.3% of the sulfur in the pyrite form, 51.0% of the organic form, and 43.8% of the sulfide form were successfully removed. The treatment was successful in reducing ash content by 33% as well. In conclusion, 65.4% of the total sulfur was eliminated by bio-desulfurization. Currently, describes a study on the biodesulfurization of coal from Indonesia by five microorganisms. Citrobacter murliniae enabled the highest observed percentage of total sulfur removal, which was 19.61%; nevertheless, it was least successful in reducing the ash content. For better outcomes in terms of total sulfur removal and ash removal, the bacterial consortium may be taken into account. Recent research shows the

Thiobacillus-induced sulfur reduction from petcoke collected from Gujarat, India. After 4 weeks of treatment, the acidophilic bacterium was able to eliminate more than 92% of the total sulfur. It has been shown that the bacterial consortium consisting of *Thiobacillus thiooxidans* and *Thiobacillus ferrooxidans* is more efficient at desulfurizing than using a pure culture of a single bacterial species. The environmentally relevant and industrially significant process of microbe-facilitated reduction in distinct sulfur forms present as impurity in diverse coal types is necessary for its safe exploitation in order to produce electricity without harming the ecosystem. The capacity of distinct strains of *Rhodococcus erythropolis* and *Rhodococcus qingshengii* developing and resting cells in sulfur exclusion has been demonstrated by Etemadifar et al. In media with no carbon source, actively proliferating cells of strain PD1 had a maximum sulfur removal rate of 46% compared to resting cells. Further evidence that strain PD1 is suitable for industrial coal cleanup procedures comes from the fact that both resting and proliferating cells of the strain were more active in removing sulfur than other strains under consideration. All of the chosen strains could also desulfurize DBT through the 4S route. The industrialization of the method for effective sulfur removal may benefit from the results of the pilot scale experiment. The industrial system to address the issue of coal burning-induced atmospheric sulfur dioxide (SO_2) pollution would benefit from being designed with the help of the full investigation of the process under laboratory conditions. Table 3.1 lists some of the significant bacterial species that contribute to desulfurization.

3.3.1 Heterotrophic Bacteria

Numerous heterotrophic microorganisms have been identified in the literature to be capable of degrading DBT. The discussion of current data on DBT biodegradation by heterotrophs continues below. The heterotrophic bacteria that are involved in the biodegradation of DBT by attacking its sulfur component are listed in Table 3.1. It has been discovered that thermophilic and mesophilic bacterial strains, including *Klebsiella* sp. 13 T and Pantoea agglomerans, breakdown DBT derived from various petroleum types. *Sphingomonas subarctica* T7b and *Lysinibacillus sphaericus* DMT-7 are two gram-negative organisms that have recently been discovered to have the ability to desulfurize DBT and its respective alkylated derivatives. It has also been noted that *Spingomonas* sp. strain XLDN2–5 may break down dibenzofuran and dibenzothiophene when carbazole is present as a substrate. Around 90% of DBT could be broken down by the microbe after 40 hours of incubation via the ring cleavage and sulfo-oxidation pathways. Actinobacteria *Oerskovia* sp., found in soil, have reportedly just been discovered to have the capacity to function as a biocatalyst for DBT desulfurization. Additionally, it has recently been revealed that freshly obtained gram-positive *R. erythropolis* SHT87 strains from oil-contaminated soils in Iran breakdown DBT present in aqueous and biphasic medium within 10 h of incubation. Additionally to DBT, it was discovered that the bacterial isolate SHT87 could use DBT sulfone, thiophene, 2-methylthiophene, and dimethylsulfoxide as its

Table 3.1 Overview of bacterial species for sulfur removal

Bacterium	Growing/ resting state	Sulfur substrate	End product	Organic phase	Temp. (C)	Reaction time (h)	Reference
Strain RIPI-S81	Growing cells	DBT	2-HBP	e	30	96	[7]
		4-MDBT	2-hydroxydimethylbiphenyl				
		4,6-DMDBT	2-hydroxy-3′-methyl-biphenyl 2-hydroxy-3-methyl-biphenyl				
Bacillus subtilis WU-S2B	Growing cells	DBT 2,8-DMDBT 4,6-DMDBT 3,4-benzo DBT	Corresponding monohydroxy biphenyls	e	50	120	[8]
Paenibacillus sp. A11–2	Growing cells	4-MDBT 2,8-DMDBT 3,4,6-TMDBT Ethyl DBT Propyl DBT	Corresponding monohydroxy biphenyls	n-tetradecane	50	22	[9]
Rhodococcus erythropolis H-2	Resting cells	3,4-benzo DBT	a-hydroxy-b-phenylnaphthalene	e	30	12	[10]
		2,8-DMDBT 4,6-DMDBT	Monohydroxy dimethyl biphenyls				
Rhodococcus sp. ECRD-1	Growing cells	DBT	HBP	e	25	12	[11]
		4,6-DEDBT	Corresponding hydroxydiethyl biphenyl				
Rhodococcus erythropolis XP	Growing cells	DBT	2-HBP	e	30	42	[12]

		4-MDBT	2-hydoxy-30-methyl-biphenyl 2-hydroxy-3-methyl-biphenyl				
		4,6-DMDBT BNT	2-hydroxy-3,30-dimethyl-biphenyl				
		BNT	a-hydroxy-b-phenyl-naphthalene				
Mycobacterium sp. MR65	Growing cells	10-methyl BNT	1-methoxy-2-(3-methylphenyl) naphthalene 1-hydroxy-2-(3-methylphenyl)naphthalene 2-(2-methoxy-3-methylphenyl) naphthalene 2-(2-hydroxy-3-methylphenyl)naphthalene	n-hexadecane	30	120	[13]
Mycobacterium phlei WU-F1	Resting cells	2-ethylNTH	2-ethylNTH sulfoxide 1-(20-hydroxynaphthyl)-1-butene 1-naphthyl-2-hydroxy-1-butene	n-tridecane	50	8	[14]

only sources of sulfur for growth. Similar to this, it has been reported that *Shewanella putrefaciens* NCIMB8768 can produce up to 43.5 mmol/L 2-HBP in 72 h when given 0.3 mM DBT. *Microbacterium* sp. NISOC-06, a new strain, desulfurized 1 mM/L of DBT to nearly 95% of its original sulfur content in under 2 weeks. Three newly discovered strains of *Rhodococcus* sp. and *Stenotrophomonas* sp., designated SA11, SA21, and SA31, respectively, were recently discovered in oil-contaminated soil on Abu Ali Island in the Eastern Province of Saudi Arabia. These strains had DBT biodegradation activity that produced 2-HBP at a rate of 4 mol/g dry cell weight/h and reached a maximum rate of 11 mol/g dry cell weight/h. The deterioration rates for SA21 and SA31, in contrast to SA11, were, however, around 20% slower. The dszA, dszB, and dszC genes were validated by molecular analyses of the plasmid containing the dsz operon, demonstrating that the microbial strains followed the 4S pathway of DBT biodegradation. Some bacterial cultures have been examined for their capacity to eliminate this source of sulfur in kerosene because DBT is a significant source of sulfur in petroleum. Within 72 h of incubation, *Desulfobacterium indolicum*, one of these cultures, reduced the sulfur content of a kerosene sample from 48.68 to 13.76 ppm. Another kerosene sample that contained sulfur was taken from the Iranian refinery in Isfahan, and it was demonstrated that a native bacterial strain, *Bacillus cereus*, could eliminate 33% of the sulfur from the sample in 72 h. The degradation of DBT, carbazol, and dibenzofuran in coking water by *Arthrobacter* sp. W1 was investigated in magnetic gellan gum. The strain that was used in this study was already naphthalene-grown. In addition to DBT, carbazole, and dibenzofuran, the magnetically immobilized cells were efficient biocatalysts for the breakdown of other hazardous chemical substances (discussed in detail in Sect. 3.3). Pan et al. attempted to improve the effectiveness of microbial desulfurization using a unique directed evolution strategy in a distinct investigation. Obtaining biocatalysts with increased desulfurization activity was the goal. *Rhodococcus opacus* was consistently cultivated in a medium with DBT as the only sulfate source. By modifying the dsz operon to make sulfur-rich polypeptides easier to encode, the desulfurizing microbe was in this case designed to demonstrate increased desulphurization. The *R. opacus* strains expressing dszABC and dszAS1BC were found to be capable of demonstrating a 20-fold increase in specific desulfurization rate after 40 transfers, which was higher than that of *R. erythropolis* IGTS8. Previous investigations into the site of the cells' biodesulfurization activity revealed that it took place intracellularly. The soluble version of the DBT biodegradation pathway's enzymes has been discovered in the cytoplasm of *R. erythropolis* IGTS8. However, there is currently no major evidence available addressing the intracellular absorption of DBT. While the enzyme catalyzing biodesulfurization was described as being a member of the cell membrane, other researchers have hypothesized that in the strain IGTS8, the sulfur removal activity is related with the cell surface. The desulfurization activity is thought to occur in the cell membrane, which is easily accessible to both the substrate and the cofactors, according to the literature currently available.

3.3.2 Fungal Strains

In addition to the bacterial strains indicated above, a number of fungal strains have also been linked to the biodesulfurization of DBT and other aromatic compounds. These strains are included in Table. Through the development of extracellular enzymes with a very high specificity and their cytochrome P-450, they have the advantage of working on a variety of aromatic chemicals. *Exophiala spinifera* strain FM, a fungus, was recently tested for its capacity to degrade DBT. It was shown that after 7 days of incubation, the fungus was able to desulfurize 0.3 mM DBT to 99 percent of its original state. It had a comparable route to the 4S process and was capable of producing 2-hydroxy-biphenyl (2-HBP). But it had the capacity to further break down 2-HBP into 1,3-benzenediol, 5-hexyl, which could lessen its damaging effects on living microbial cells. The capacity of *C. elegans* to break down DBT into DBT-5-oxide and DBT-5-dioxide has also been documented. However, it was discovered that *C. elegans* did not produce any biphenyl. Agaric basidiomycetes like *Agrocybe aegerita* and *Coprinellus radians* have also been investigated to see how well they degrade DBT in vivo and in vitro by biodesulfurization. It was found that *A. aegerita* could completely oxidize DBT within 16 days of incubation, resulting in eight different metabolites, including ring-hydroxylation products, DBT sulfoxide, and DBT sulfone. It was discovered that 60% of DBT in the case of *C. radians* was transformed, resulting in the creation of DBT sulfone and DBT sulfoxide. Findings suggested that DBT might be degraded by whole cell and purified extracellular peroxygenases. Even under natural circumstances, peroxygenases might have contributed to the oxidation of DBT. Some fungal strains have the potential to desulfurize coal in addition to aromatic chemicals. *Paecilomyces* TLi has previously been shown to break down coal-related chemicals. Through the sulfate oxidation pathway, it was able to break down dibenzothiophene, producing 2,2′-dihydroxybiphenyl in the process. In addition to DBT, the fungus was able to convert diphenyl sulfide and ethyl phenyl sulfide into the appropriate sulfone. *Trametes versicolor*, *Phanerochaete chrysosporium*, and *Pleurotus sajor-caju* are three different fungal strains that have been utilized to treat subbituminous coal from the Pirin Basin of Bulgaria that contains a lot of sulfur. According to the study's findings, a mixed bacterial culture was able to remove about 79 percent of inorganic sulfate, while the fungal strains *T. versicolor* and *P. chrysosporium* were able to further lower the organic sulfate contained in the coal sample by 13%. In one of the investigations, under ideal circumstances, laccase enzyme generated from *T. versicolor* ATCC 200801 could lower the overall sulfur content of Turkish lignites by 29%, of which 35% was pyritic, 13% sulphate, and 25% organic. The native *Aspergillus* sp. has also been tested for its capacity to desulfurize Assam coal produced from India's North Eastern Coalfields. An employed size fraction for the desulfurization study was 74 m. Within 10 days of incubation, 70–80 percent of the total sulfur was removed as a result of reducing the coal's sulfur content from 2.13 to 0.47%. The findings indicated that the fungus targeted organic sulfur in particular because it made up a large portion of Assam coal. For the optimal decomposition of model lignite compounds, research has been done on the white rot fungi, *Coriolus versicolor*, and golden mushroom. The model chemicals were vulnerable to assault by the fungus. Gas chromatography and IR analyses revealed that several degradation products were created during the breakdown of the model compounds. For coal degradation, the combined effects of two deuteromycetes, *Trichoderma atroviride* and *Fusarium oxysporum*, have been investigated. Three steps were used to describe the fungal strains' course of action: First is the production of alkaline substances through *Fusarium* oxysporum, and then second is the action of chelators produced during growth on glutamate-containing media. The last stage involved *T. atroviride* secreting hydrolytic enzymes that helped to solubilize the coal and break down humic acids. The pretreated coal sample (processed with HNO3) was attacked by a natural fungus strain, *Paecilomyces* TLi, in a sequential coal solubilization investigation conducted by Davison et al. After being treated with a methanogen, the first step's byproducts produced biogas and low molecular weight molecules. Instead of using the fungal strains *Penicillium chrysogenum* and *Aspergillus fumigatus*, Sharma et al. used the bacterial strains *Escherichia coli* and *Streptococcus* sp. According to the study's findings, fungal strains might destroy coal more effectively than bacterial ones. In yet another investigation, *T. versicolor* ATCC 200801 and *P. chrysosporium* ME446 of the white rot fungus were used to biodesulfurize Turkish lignites. Within 6 days of incubation, a maximum sulfur reduction of 40% may be accomplished under ideal circumstances. The investigation revealed that coal's calorific value remained unaltered. During the procedure, the ash and metal impurities were eliminated. The natural fungus *Stachybotrys* sp. WS4 proved successful in selectively removing 76 and 64.8% of the sulfur from heavy crude oil recovered from the Soroush and Kuhemond oil fields, respectively, within 72 and 144 h. The ability of the fungal strains to remove sulfur was also contrasted with that of BBRC-9016, a thermophilic bacterial strain. It was discovered that the former was superior to the bacterial strain in terms of effectiveness. Similar biodesulfurization ability of the fungus *C. elegans* var. echinulata, which can break down benzothiophene together with methyl, ethyl, and diethyl-benzothiophene to create sulfoxide molecules, has been demonstrated in studies by Matsui et al. The biodesulfurization mechanism heavily depends on the bulk transfer of organic molecules. Keeping this in mind, the white rot fungus *Bjerkandera* sp. BOS55 was used in a slurry phase stirred tank reactor to biodesulfurize dibenzothiophene, fluoranthene, pyrene, and chrysene from wetland soil. The fungal strain targets the aromatic molecules, changing their structure and causing them to degrade through the secretion of ligninolytic enzymes. In the presence of soil microflora, the breakdown of the organic compounds was slower; nevertheless, after 30 days of incubation, the degradation increased to 30 mg PAH/kg soil.

3.3.3 *Yeasts*

Other than the fungi and heterotrophic bacteria, several yeasts can also encourage biodesulfurization. One of the earliest strains of yeast to be reported to have bidesulfurization activity is *Rhodosporidium toruloides* DBVPG 6662. The study's findings demonstrated that 27% of the sulfur in DBT could be removed after 36 h of incubation when glucose was present. Within 15 days of incubation, the strain was able to extract 68% of benzo- and dibenzothiophene from a bitumen emulsion and 11% of total organic sulfur from fuel oil that wasn't emulsified. These findings demonstrated that this yeast strain had the ability to remove organic sulfate from fuel oils and sulfate-containing substances. There have also been reports of yeasts like *Trichosporon* sp. and *Candida* sp. being able to encourage biodesulfurization. Tests have been done to see if *Trichosporon* sp. can desulfurize coal. It was discovered that the microbial strain could break down DBT almost entirely within 14 days of incubation when glucose wasn't present. It might concurrently affect biphenyl and avoid the medium from being too contaminated with the microbial inhibitor biphenyl. Therefore, this strain can break down both harmful substances, namely, DBT and biphenyl.

3.4 Microorganisms for Inorganic Sulfur Removal

3.4.1 *Obligate Autotrophs*

Different strains of *Thiobacillus ferrooxidans* are the most often employed obligatory autotrophs for coal desulfurization [29–31], and nearly all strains have been obtained from acid mine fluids. Gram-negative rods (0.5 by 1–2 tm), flagellated, non-spore-forming, acidophilic (pH optimum, 2–3), mesophilic (temperature optimum, 25–30 °C), aerobic, and flagellated are the characteristics of these bacteria. Taxonomically speaking (such as by G/C composition), the different strains have not been distinguished. The Calvin cycle is used by the organisms to fix COz when they are growing in an autotrophic environment, and they use the oxidation of reduced sulfur and iron compounds as a source of energy (Fig. 3.1). Under autotrophic circumstances, the organisms have been employed to leach metals (Cu, Zn, U, Ni) from low-grade sulfide ores and remove pyritic sulfur from coal 2′2°′21. Other *Thiobacillus* species, such as *T. perometabolis* and *T. organoparus*, can be utilized to remove pyritic sulfur from coal by oxidizing reduced sulfur and iron compounds. While the obligately autotrophic *T. ferroxidase* strains are poisonous to some organic molecules (such as agar), other *Thiobacillus* species can metabolize some organic chemicals [31]. Although there are some *Thiobacillus*-type organisms that metabolize dimethyl sulfide and dimethyl disulfide [29,33], neither these nor the nitrogen-fixing *Thiobacillus* species 23 have been tested for coal desulfurization. Instead, it is thought that *Thiobacillus* species are incapable of removing organic

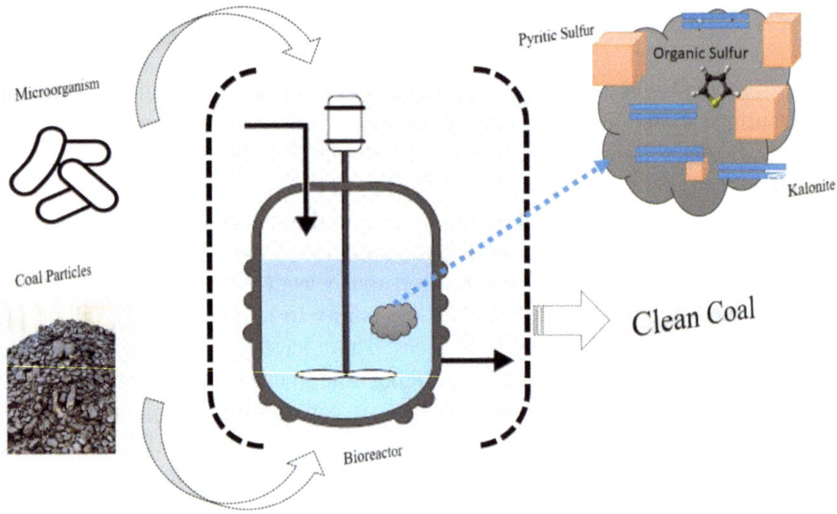

Fig. 3.1 Role of Bacteria for sulfur removal

sulfur from coal [21]. When coal particle sizes are tiny enough (diameter 80 l m), *Thiobacillus* species may remove more than 90% of the initial pyritic sulfur; nonetheless, the main issue with utilizing mesophilic species is the slow rate of removal (residence times of 4–5 days are required in continuous operations). Thermophilic organisms (temperature optimum, 50–60 °C) are some of the autotrophic species that can remove pyritic sulfur from coal. Organisms of the *Thiobacillus* type (TH-bacteria) have been found in sulfur-rich soil or acidic hot springs [15, 34, 35]. For these organisms to develop, certain organic substances (such as yeast extract or cysteine) are necessary. The removal of pyritic sulfur from coal was accomplished using thermophilic *Thiobacillus*-type organisms (temperature optimum, 55 °C), and it is feasible that thermophilic strains could speed up the process. However, their most significant contribution might be in enabling reactor operation at temperatures where the likelihood of contamination is reduced.

3.4.2 Thermophilic Facultative Autotrophs

Reduced iron and sulfur compounds can be oxidized by a number of facultative autotrophic species found in Yellowstone National Park's acidic hot springs: For example, Sulfolobus brierleyi, an organism that thrives in thermophilic environments (50–60 °C) and is facultatively autotrophic [36], may be used to extract pyritic sulfur and maybe organic sulfur from coal, though this has not been tested. Adding yeast extract to the nutritional media helped this organism grow more quickly on ferrous iron and elemental sulfur. *Sulfolobus acidocaldarius* is a thermophilic (60–90 °C), acidophilic (pH, 1.5–4), and facultative autotrophic organism

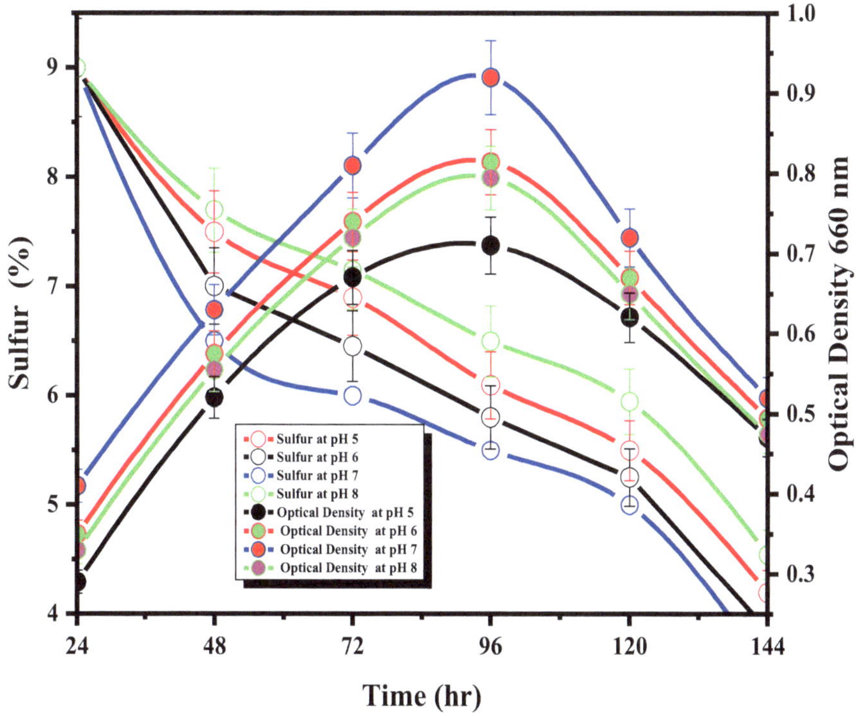

Fig. 3.2 Effect of different pH ranges on growth of cells and sulfur reduction using *Rhodococcus* sp. SL-9 on Lakhra coal along with other parameters constant 30 °C, 170 rpm, and 10% glucose concentration

that can oxidize reduced iron and sulfur [37, 38]. It can also use simple organic compounds as its energy source, such as glucose and yeast extract [38]. The organism lacks a solid cell wall, which is typical of the class Archaebacterial (i.e., it has no peptidoglycan layer in its cell wall). It could oxidize dibenzothiophene and was frequently used to remove pyritic and organic sulfur from coal [24–28]. *S. acidocaldarius* is depicted in suspension culture in Fig. 3.2 and is captured on a coal particle surface in Fig. 3.3 in a scanning electron microscope (SEM) image. Cell affixation SEM electron probe analysis revealed the presence of iron and sulfur at the attachment point, indicating that *S. acidocaldarius* attached itself only to pyrite surfaces in coal (Kargi, F., unpublished results). No pertinent extracellular enzymes were found in the culture media, indicating that cell attachment to pyrite surfaces in coal is crucial for pyrite oxidation. It is not apparent, nevertheless, whether organic sulfur elimination necessitates cell attachment. In general, cell attachment depends on the composition of the nutritive medium and the hydrodynamics of the reactor as well as the surface characteristics of the coal/pyrite and the organisms. To get the best cell attachment, it is important to adjust the cell density at the coal's surface. It may be possible to increase the rate of sulfur removal by having a fundamental grasp of the mechanisms governing cell attachment in coal desulfurization. A new bacterium called *Sulfobacillus thermosulfidooxidans* was recently discovered in Russia [41, 42], and several *Sulfolobus*-like species have been recovered from hot sulfatara regions in Italy and Japan [39, 40]. Although there haven't been any reports of this rod-shaped, thermophilic, acidophilic, facultative autotrophic, or spore-forming bacterium being used to desulfurize coal or oxidize model organic sulfur compounds, it may oxidize reduced iron and sulfur compounds. The following are benefits of thermophilic and facultative autotrophic species (such *Sulfolobus* spp.) over mesophilic and obligate autotrophic organisms (e.g. *Thiobacillus* spp.).

Operations at high temperatures decrease the possibility of growth media contamination.

Coal can have both pyritic and organic sulfur removed by facultative autotrophs.

- The system can run at high cell and coal concentrations at high temperatures without incurring cooling costs.
- When operating at high temperatures (50–80 °C), pyritic sulfur is chemically oxidized at a faster rate by ferric iron, which is created when pyritic iron in coal is oxidized by microorganisms.

3.4.3 Heterotrophs Filamentous Fungi

Several filamentous fungi have the ability to break down some organic sulfur compounds. Sulfate ions are released into the medium when certain fungi's sulfatases catalyze the oxidation of sulfonated phenol molecules [43]. This implies that they might be successful in eliminating organic sulfur from coal even if no such reports have been made public. A number of fungi, including yeasts and molds, have recently been identified from coal disposal sites and used to solubilize coal [44]. These organisms can potentially be helpful for coal desulfurization since they oxidize some aromatic chemicals and turn them into soluble polar molecules. Organosulfur removal from oil has been accomplished using a number of isolated pseudomonad species [45–46]. In about 6–10 h, these organisms can oxidize dibenzothiophene and remove a sizable amount of organic sulfur from oil. Some of these species (such *Pseudomonas alcaligenes*) have undergone genetic modification, which has enhanced their capabilities. Oil desulfurizing organisms may be employed to desulfurize coal, particularly coal that has been liquified or solubilized. According to Atlantic Research Company, a different strain of *Pseudomonas* could oxidize dibenzothiophene and remove 20–30% of the organic sulfur from coal in 10–12 h. Mixed cultures of *T. ferrooxidans* and *T. thiooxidans*, an acidophilic and autotrophic organism, outperformed pure cultures of *T. ferrooxidans* in the removal of pyritic sulfur from coal [22, 23]. The function of *T. thiooxidans* (in the mixed culture) is to oxidize elemental sulfur, which is created when *T. ferrooxidans* oxidizes pyritic sulfur to sulfate. Similar to this, pyritic sulfur removal from c0al14 can also be accomplished using a combined culture of *T. thiooxidans* and *Leptospirillum ferrooxidans*: Ferrous iron can be oxidized by *L. ferrooxidans* to ferric iron, which in turn oxidizes pyritic sulfide to elemental sulfur before being further oxidized by *T. thiooxidans* to sulphate. The reduction of pyritic sulfur from coal was accomplished using a different, more ambiguous mixed culture of organisms found in acidic coal mine streams [47].

3.4.3.1 Effect of Parameters on Sulfur Removal from Coal

Microbial desulfurization, also known as biodesulfurization, is a biological approach to remove sulfur from coal using microorganisms. Several parameters can influence the efficiency of this process.

- The choice of microorganisms is critical. Sulfur-reducing bacteria, such as *Thiobacillus* spp., can oxidize sulfur compounds in coal to less harmful forms. The selection of the most suitable microorganism depends on the type of sulfur compounds present in the coal and the operating conditions.
- The type and concentration of sulfur compounds in the coal can vary widely. Microorganisms may have preferences for specific sulfur compounds, so understanding the sulfur composition of the coal is essential for optimizing the process.
- Microorganisms require specific nutrients, such as carbon sources, nitrogen, phosphorus, and trace elements, for growth and metabolism. Adequate nutrient supply is crucial to support microbial activity and sulfur removal.
- The pH of the system can affect microbial activity. Different microorganisms have different pH preferences, so maintaining the appropriate pH range is important for optimal desulfurization.
- Temperature can significantly influence microbial growth and activity. Each microorganism has an optimal temperature range, and the process should be operated within this range to maximize efficiency.
- Some sulfur-reducing bacteria are facultative anaerobes, meaning they can operate in the presence or absence of oxygen. The availability of oxygen or the absence of oxygen can impact the microbial metabolism and sulfur removal rates.
- The size of coal particles can affect microbial access to sulfur compounds. Smaller particle sizes can provide more surface area for microbial attachment and enhance desulfurization.
- The time coal spends in contact with the microorganisms can impact the extent of sulfur removal. Longer residence times may allow for more complete desulfurization.
- Microorganisms often form biofilms on coal particles or in reactors. Biofilms can provide a protective environment for microbes and enhance their desulfurization capabilities.
- Some substances in coal or the environment can inhibit microbial activity. Understanding potential inhibitors and managing their presence are crucial for successful biodesulfurization.

- Regular monitoring of microbial activity, sulfur levels, and other relevant parameters is essential for process control and optimization.
- Scaling up biodesulfurization processes from laboratory-scale to industrial-scale operations requires careful consideration of reactor design and engineering factors.

It's important to note that biodesulfurization is typically more effective at removing organic sulfur compounds from coal rather than inorganic forms like pyritic sulfur. Additionally, the success of microbial desulfurization depends on the specific conditions and microorganisms used, so research and optimization are necessary to tailor the process to a particular coal feedstock and desired sulfur removal goals.

3.4.3.2 Effect of pH

The pH of the environment can have a significant impact on the activity of *Rhodococcus* species in organic sulfur removal. *Rhodococcus* species are known to produce sulfur-specific enzymes, such as sulfur oxygenase reductase (SOR), that are responsible for the oxidation of organic sulfur compounds. The activity of these enzymes can be influenced by the pH of the environment. In general, *Rhodococcus* species prefer a slightly alkaline pH range of 7.0–8.0 for optimal organic sulfur removal. This is because the SOR enzyme is most active in this pH range, and the alkaline conditions can also promote the growth and activity of the bacteria. However, the specific optimal pH for organic sulfur removal by *Rhodococcus* species can vary depending on the strain and the type of sulfur compound being targeted. For example, some *Rhodococcus* strains have been shown to have a pH optimum closer to neutral pH, around 6.5–7.5, for the removal of certain organic sulfur compounds. It is important to note that extreme pH values, such as highly acidic or highly alkaline conditions, can inhibit the activity of *Rhodococcus* species and reduce their ability to remove organic sulfur compounds. Therefore, it is crucial to maintain an appropriate pH range for optimal bacterial activity. In summary, the pH of the environment can have a significant impact on the activity of *Rhodococcus* species in organic sulfur removal. In general, a slightly alkaline pH range of 7.0–8.0 is preferred, but the optimal pH can vary depending on the specific strain and type of organic sulfur compound. Maintaining an appropriate pH range is essential for optimal bacterial activity and effective organic sulfur removal (Fig. 3.3).

In the quest to pinpoint the optimal pH for both cell growth and desulfurization activity of *Rhodococcus* spp. (SL-9), a profound correlation between robust cell growth and enhanced desulfurization activity was established at a pH of 7.0. This observation is of particular significance since it was noted that the desulfurization activity of isolate SL-9 is intricately linked to its growth rate. By maintaining the pH at an initial value of 7.0, a notable achievement was reached: approximately 0.19 ± 0.03 millimoles (mM) of dibenzothiophene (DBT) underwent desulfurization within a span of 72 hours. Concurrently, the cell concentration exhibited impressive growth, reaching 1.8 ± 0.1 grams per liter (g/L). These outcomes

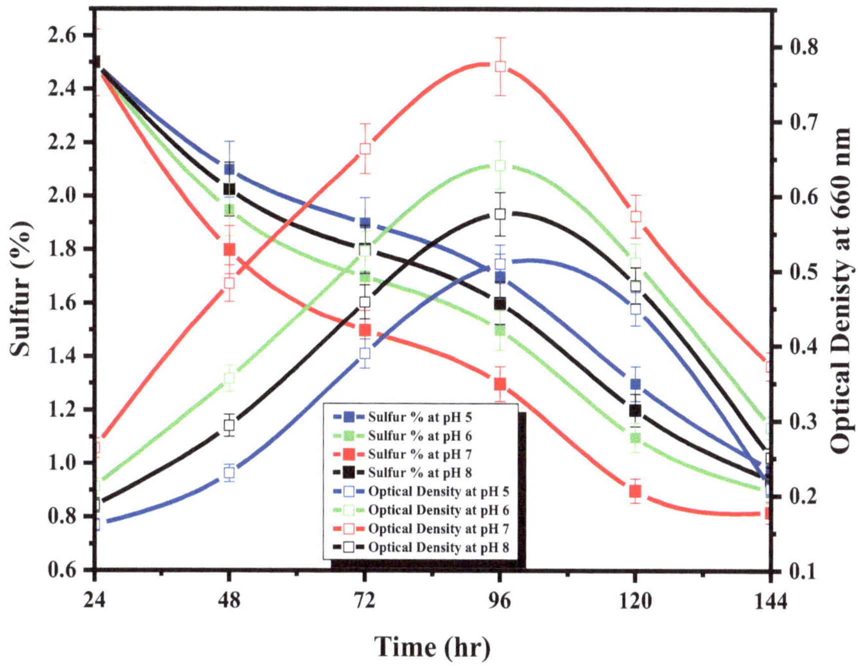

Fig. 3.3 Effect of different pH ranges on growth of cells and sulfur reduction using *Rhodococcus* sp. SL-9 on Thar coal along with other parameters constant 30 °C, 170 rpm, and 10% glucose concentration

underscore the pivotal role that pH optimization plays in fostering both the growth of *Rhodococcus* spp. (SL-9) and its capacity for effective desulfurization. It's worth mentioning that throughout the growth process, a reduction in pH value was also observed. This pH drop might be attributed to the metabolic activities of the microorganisms, which can lead to the generation of acidic byproducts. Overall, this investigation highlights the significance of maintaining an optimal pH environment to cultivate thriving microbial populations with heightened desulfurization capabilities.

3.5 Future Challenges

It has been extremely difficult to design a culture with highly specialized coal-desulfurizing capabilities [9]. The research's other primary obstacle is increasing the rate of biodesulfurization of a variety of molecules that include sulfur. The activity of the biocatalysts now used for desulfurization must be multiplied by 500 [33]. Numerous studies have been conducted in this area using genetic engineering to enhance the secretion of microbial enzymes by optimizing the concentration of all four enzymes and determining the necessity of upping the level of Dsz D enzyme while concurrently removing Dsz B enzyme [176]. Higher desulfurization efficiency at a reduced cost can be achieved by successfully removing the microbial cells from the sample and reusing them later. Even while the cell immobilization method and the use of nanoparticles have made some progress in this field, more research could help achieve higher desulfurization rates. It's also important to increase the cells' lifespan. The enzymes that were depleted or harmed during the reaction process can then be recreated by the cells [63]. To aid in the biodesulfurization of crude oil fractions, highly resistant strains with high solvent tolerance capacity and the capacity to function in the presence of nonaqueous solvents should be created. Purified enzymes and proteins can be utilized instead of living cells because they are highly effective and don't have the issues that come with using living cells. Surfactants also facilitate the mass transfer of organic molecules from the organic to the aqueous phases, which suggests that their use may accelerate biodesulfurization. But first, it's important to assess the surfactant's toxicity against microbes and figure out the ideal surfactant dosage for effective biodesulfurization. New insights into the topic of biodesulfurization can be gained by having a better and deeper grasp of all these elements.

3.6 Conclusion

Due to their environmental friendliness and cheap operating costs, microbial systems for biodesulfurization are gradually gaining more appeal. The number of microbial strains that have been identified is enormous, and researchers are always working to increase the rate of biodesulfurization, which has the potential to become the most effective method for removing sulfur for a very long time. The ability of genetically engineered strains to tolerate solvents and remove sulfur is being researched and tested. Immobilization and adsorption approaches are being used to improve catalytic activity while keeping cost and desulphurization efficiency in mind. In this procedure, the microbial cells are recycled multiple times. The development of a more environmentally friendly sulfur removal process will be aided by a deeper comprehension of the fundamental mechanisms underlying microbial action, exploration of additional desulfurization pathways, and the establishment of optimal parameters for large-scale biodesulfurization.

References

1. Singh VK, Singh R, Kumar A, Bhadouria R, Notarte KI (2021) Perspectives in desulfurization of coal using microbes. In: Microbe mediated remediation of environmental contaminants. Elsevier, pp 141–155

2. Jatoi AS, Aziz S, Soomro SA (2021) Effect of native microorganism Rhodococcus spp. SL-9 for dibenzothiophene degradation and its application towards clean coal approach. Cleaner Eng Technol 3:100126
3. Jatoi AS, Aziz S, Soomro SA (2021) Experimental and numerical investigation of DBT degradation via Rhodococcus spp.(SL-9) through the use of biological assisted method. Biomass Convers Biorefinery 13:1–9
4. Jatoi AS, Hashmi Z, Anjum A, Bhatti ZA, Siyal SH, Mazari S et al (2021) Overview of bioelectrochemical approaches for sulfur reduction: current and future perspectives. Biomass Convers Biorefinery 13:1–16
5. Jatoi AS, Aziz S, Soomro SA (2024) Bioprocess development and optimization of parameters for enhancing bio-assisted method for sulfur removal from Lakhra coal using Rhodococcus sp SL-9. Int J Coal Preparation Util 44:37–50
6. Jatoi AS, Aziz S, Soomro SA (2021) Biological assisted organic sulfur removal from low rank indigenous coal using airlift bioreactor. Bioprocess Biosyst Eng 44:417–427
7. Rashidi L, Mohebali G, Darian JT, Rasekh B (2006) Biodesulfurization of dibenzothiophene and its alkylated derivatives through the sulfur-specific pathway by the bacterium RIPI-S81. Afr J Biotechnol 5:351–356
8. Kirimura K, Furuya T, Nishii Y, Ishii Y, Kino K, Usami S (2001) Biodesulfurization of dibenzothiophene and its derivatives through the selective cleavage of carbon-sulfur bonds by a moderately thermophilic bacterim Bacillus subtilis WU-S2B. J Biosci Bioeng 91:262–266
9. Ishii Y, Konishi J, Okada H, Hirasawa K, Onaka T, Suzuki M (2000) Operon structure and functional analysis of the genes encoding thermophilic desulfurizing enzymes of Paenibacillus sp. A11-2. Biochem Biophys Res Commun 270:81–88
10. Ohshiro T, Hirata T, Izumi Y (1996) Desulfurization of dibenzothiophene derivatives by whole cells of Rhodococcus erythropolis H-2. FEMS Microbiol Lett 142:65–70
11. Grossman M, Lee M, Prince R, Minak-Bernero V, George G, Pickering I (2001) Deep desulfurization of extensively hydrodesulfurized middle distillate oil by Rhodococcus sp. strain ECRD-1. Appl Environ Microbiol 67:1949–1952
12. Tao F, Zhao P, Li Q, Su F, Yu B, Ma C et al (2011) Genome sequence of Rhodococcus erythropolis XP, a biodesulfurizing bacterium with industrial potential. Am Soc Microbiol 193:6422
13. Chen H, Zhang W-J, Chen J-M, Cai Y-B, Li W (2008) Desulfurization of various organic sulfur compounds and the mixture of DBT+ 4, 6-DMDBT by Mycobacterium sp. ZD-19. Bioresour Technol 99:3630–3634
14. Furuya T, Kirimura K, Kino K, Usami S (2001) Thermophilic biodesulfurization of dibenzothiophene and its derivatives by Mycobacterium phlei WU-F1. FEMS Microbiol Lett 204:129–133

Chapter 4
Sulfur Removal Mechanism Using Microorganisms

Abstract Biodesulfurization (BDS) has emerged as a sustainable and eco-friendly alternative to conventional sulfur removal methods for fossil fuels. This process utilizes microorganisms that leverage specific metabolic pathways, such as the 4S and Kodama pathways, to selectively break down sulfur-containing compounds like dibenzothiophene (DBT) while maintaining the integrity of the hydrocarbon structure. The 4S pathway involves a series of enzymatic reactions mediated by DszA, DszB, and DszC, converting DBT into non-toxic products. In contrast, the Kodama pathway employs different enzymes, resulting in the complete mineralization of sulfur.

Although BDS offers significant advantages, its industrial application is limited by challenges such as slow reaction rates, substrate specificity, and the presence of inhibitory substances. Recent advancements in biotechnology, particularly genetic and metabolic engineering, are focused on improving the efficiency and resilience of desulfurizing microorganisms and their enzymes. Additionally, the use of omics technologies has provided valuable insights into microbial metabolic pathways and regulatory mechanisms, enabling the development of customized strains with enhanced desulfurization performance. Exploring alternatives like non-cellular biocatalysts and multi-enzyme systems also presents promising strategies to boost the efficiency of BDS processes.

This chapter highlights the current state of research in biodesulfurization mechanisms and emphasizes the importance of overcoming existing challenges to advance the application of BDS technologies. As industries increasingly seek sustainable solutions to meet stringent environmental regulations, the continued exploration of innovative biotechnological approaches will be crucial for developing efficient methods for sulfur removal, ultimately contributing to cleaner fuel production and reducing the environmental impact of fossil fuel consumption.

A. S. Jatoi, *Biological Removal of Sulphur from Coal*, Green Energy and Technology, https://doi.org/10.1007/978-981-96-8628-5_4

4.1 Introduction

The increasing environmental concerns associated with sulfur emissions from fossil fuels have prompted the need for effective desulfurization technologies [1–3]. Even if they work well, conventional techniques like hydrodesulfurization (HDS) use a lot of energy and frequently produce hazardous byproducts. A more environmentally friendly option is biodesulfurization (BDS), which uses microorganisms' inherent metabolic processes to remove sulfur from molecules that contain sulfur in crude oil and petroleum products. In addition to addressing the problem of sulfur emissions, this biotechnological method maintains the hydrocarbon structures, minimizing the loss of important fuel constituents [4].

The 4S and Kodama pathways are the two most prominent metabolic pathways through which BDS predominantly functions. The 4S route is especially important because it uses important enzymes like DszA, DszB, and DszC to convert dibenzothiophene (DBT) into less hazardous compounds like 2-hydroxybiphenyl through a sequence of enzymatic processes. In the meantime, the Kodama process uses various enzymatic pathways to fully mineralize sulfur, producing sulfate as a byproduct [5]. Both processes demonstrate how some bacteria, including species from the genera *Gordonia* and *Rhodococcus*, may efficiently metabolize sulfur compounds in a variety of environmental settings [6].

Notwithstanding their potential, a number of obstacles prevent BDS technologies from being widely used in industry. These include the influence of inhibitory chemicals found in crude oil, poor substrate specificity, and comparatively sluggish reaction rates. However, promising approaches to improve the efficacy and efficiency of microbial desulfurization processes are provided by biotechnology breakthroughs including genetic and metabolic engineering. This introduction lays the groundwork for a more thorough examination of the processes involved in biodesulfurization, the difficulties that arise, and the opportunities for creating environmentally friendly methods of removing sulfur from the processing of fossil fuels.

4.2 Biodesulfurization

Utilizing microorganisms, biodesulfurization is a novel technique for eliminating sulfur compounds from fuels, gasses, and wastes [7, 8]. Both organic and inorganic forms of sulfur are frequently encountered, including thiophenes in crude oil and sulfates and hydrogen sulfide in wastewater and biogas. Since sulfur dioxide (SO_2) is released during combustion or processing, the presence of sulfur in fuels and industrial waste streams poses a serious environmental risk and contributes to acid rain and air pollution. In order to comply with increasingly strict environmental standards and stop industrial equipment from corroding, sulfur levels must be reduced [9].

An environmentally benign and energy-efficient substitute for conventional chemical desulfurization techniques that depend on high temperatures, pressures, and catalysts is biodesulfurization. Bacteria or other microbes that naturally break down sulfur compounds into less toxic or easier-to-manage forms are involved. The technique is more sustainable and uses less energy because these microbes can function in mild environments like room temperature and air pressure [10, 11].

The ability to handle both organic and inorganic sulfur compounds in a variety of sectors is the main benefit of biodesulfurization [12]. It can be used to desulfurize crude oil and lower emissions in the petroleum industry, as well as in biogas purification, where hydrogen sulfide is eliminated to stop corrosion and enhance fuel quality. Biodesulfurization is a crucial step in managing sulfur-related emissions in an environmentally responsible manner as companies transition to greener technologies.

4.3 Model Sulfur Compounds in BDS Studies

Sulfur compounds are usually divided into two categories in biodesulfurization (BDS) studies: organic sulfur compounds and inorganic sulfur compounds. For each to be effectively removed, a different microbial pathway is needed. The most commonly studied sulfur compounds include the following (Fig. 4.1):

4.3.1 Dibenzothiophene (DBT)

Dibenzothiophene (DBT) is a heterocyclic compound with a sulfur atom situated between two benzene rings. DBT is the primary model organic sulfur compound used in BDS studies because it is abundant in crude oil and is resistant to chemical desulfurization processes like hydrodesulfurization (HDS). It is considered the most

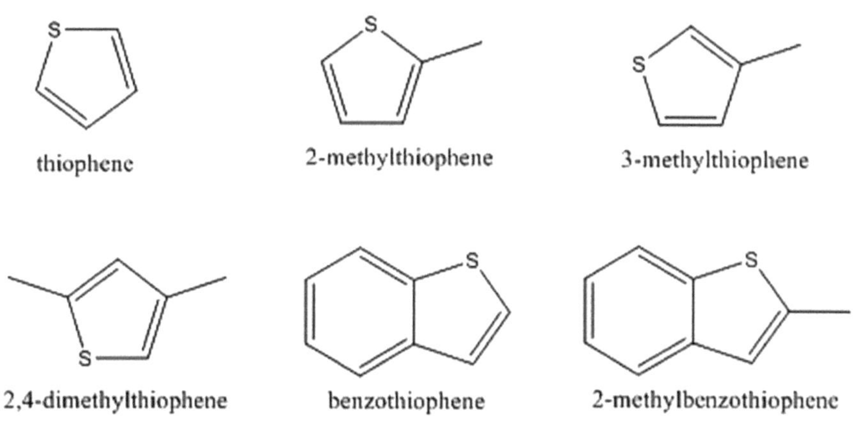

Fig. 4.1 Model sulfur compounds for BDS [13]

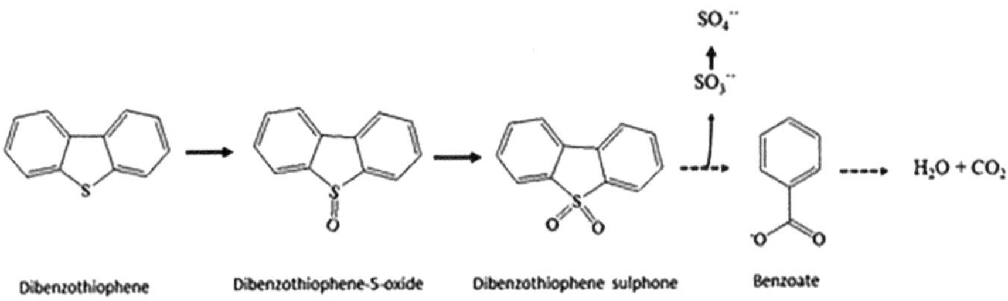

Fig. 4.2 Reaction mechanism for dibenzothiophene removal [15]

challenging sulfur-containing compound to remove, making it a benchmark for evaluating biodesulfurization efficiency [10].

The 4S pathway is the most well-known microbial degradation route for DBT. In this pathway, DBT is first oxidized to DBT-sulfoxide, then to DBT-sulfone, before desulfurization leads to the formation of 2-hydroxybiphenyl and sulfate. This pathway is employed by bacteria like *Rhodococcus* and *Gordonia*, which have been studied extensively for their ability to selectively remove sulfur without degrading the hydrocarbon framework of the molecule [14] (Fig. 4.2).

4.3.2 Thiophenes and Alkylated Thiophenes

Thiophenes are five-membered sulfur-containing heterocyclic compounds. Alkylated thiophenes (e.g., methylthiophene, benzothiophene) are thiophene derivatives where the sulfur ring is attached to alkyl groups or benzene rings. Thiophenes and their derivatives are present in varying amounts in fossil fuels, and like DBT, they are resistant to hydrodesulfurization. These compounds serve as models in BDS to understand microbial degradation of more complex organic sulfur molecules [16]. Microorganisms such as *Pseudomonas* and *Rhodococcus* strains have been shown to degrade thiophenes by breaking the sulfur bond, leading to desulfurization and producing simpler, non-sulfur aromatic hydrocarbons. The degradation of these compounds is key for cleaner fuel production.

4.3.3 Hydrogen Sulfide

Hydrogen sulfide (H_2S) is a simple, toxic, colorless gas with a characteristic rotten egg smell. It is a reduced sulfur compound [17].

H_2S is a common contaminant in biogas, natural gas, and industrial wastewater. Its removal is essential because it causes corrosion in pipelines and equipment, and it is highly toxic even in small concentrations. H_2S is a major focus in BDS because many sulfur-oxidizing bacteria (SOB) are capable of converting H_2S into elemental sulfur or sulfate.

Sulfur-oxidizing bacteria (SOB), such as *Thiobacillus* and *Acidithiobacillus*, oxidize hydrogen sulfide to elemental sulfur (S^0) or sulfate under aerobic conditions. This oxidation reaction is a crucial part of the biogas purification process and is used in several industrial applications to reduce sulfur emissions and prevent equipment damage.

4.3.4 Sulfates (SO_4^{2-})

Sulfates are inorganic sulfur compounds that consist of a sulfur atom surrounded by four oxygen atoms in a tetrahedral arrangement. Sulfates are prevalent in industrial wastewater and certain natural water sources. They are often present in high concentrations in mining, chemical manufacturing, and paper mill effluents. The reduction of sulfate is important in BDS studies aimed at wastewater treatment, as the formation of hydrogen sulfide from sulfate in anaerobic environments can cause odor issues and corrosion.

Sulfate-reducing bacteria (SRB), such as *Desulfovibrio* and *Desulfobacter*, utilize sulfate as an electron acceptor in anaerobic environments, reducing it to hydrogen sulfide (H_2S). This biological sulfate reduction is important in bioreactors and anaerobic digestion processes used for wastewater treatment.

4.3.5 Sulfur-Containing Amino Acids (Cysteine and Methionine)

Cysteine and methionine are sulfur-containing amino acids. Cysteine contains a thiol (-SH) group, while methionine contains a thioether (-S-CH_3) group [18].

These amino acids are studied as model compounds for understanding the biodegradation of sulfur-containing proteins in wastewater and organic waste. They are also involved in sulfur cycling in natural ecosystems [19].

Through a variety of metabolic processes, microorganisms break down these sulfur-containing amino acids, releasing sulfur as sulfate or hydrogen sulfide. In biological wastewater treatment, where sulfur compounds must be controlled to avoid contamination, these interactions are essential [20].

4.4 Susceptibility of DBT to Microbial Attack

Due to its great stability and resistance to traditional chemical desulfurization procedures, dibenzothiophene (DBT), a heterocyclic sulfur-containing molecule present in crude oil, is one of the most difficult sulfur compounds to remove [21]. Compared to more straightforward sulfur compounds like thiols or sulfides, its aromatic structure—which has a sulfur atom sandwiched between two benzene rings—makes it less vulnerable to microbial attack [21–23]. However, specific microorganisms have evolved pathways to target DBT, particularly through the 4S pathway, which selectively removes sulfur without degrading the hydrocarbon skeleton [24]. This pathway is a microbial strategy that allows bacteria to utilize the sulfur from DBT while preserving the energy-rich carbon backbone, making DBT an important model for studying microbial desulfurization.

The susceptibility of DBT to microbial attack is largely determined by the enzymatic machinery of the microorganisms involved [25]. Microbes like *Gordonia* and *Rhodococcus* oxidize DBT to DBT-sulfoxide in the 4S pathway, which is then further oxidized to DBT-sulfone. The elimination of sulfur is facilitated by this sequence of events, which ultimately produce 2-hydroxybiphenyl and sulfate. The remaining carbon structure remains intact, but the sulfur atom is eliminated as sulfate. Because it reduces the sulfur concentration while preserving the useful hydrocarbons, this selective desulfurization is very beneficial for fuel treatment because it minimizes the production of toxic sulfur dioxide (SO_2) during burning.

In contrast to chemical techniques like hydrodesulfurization (HDS), which necessitates high temperatures and pressures, the attack on DBT is rather gradual, even though microbial desulfurization pathways are successful. The solubility of DBT in aqueous environments, the compound's bioavailability to microbes, and the metabolic constraints of the microorganisms are some of the aspects that can prevent microbial attack on DBT. To overcome these challenges, researchers are focusing on optimizing microbial strains through genetic engineering, improving reactor designs for better mass transfer, and combining biodesulfurization with other methods to enhance the removal efficiency of DBT from fuels (Fig. 4.3).

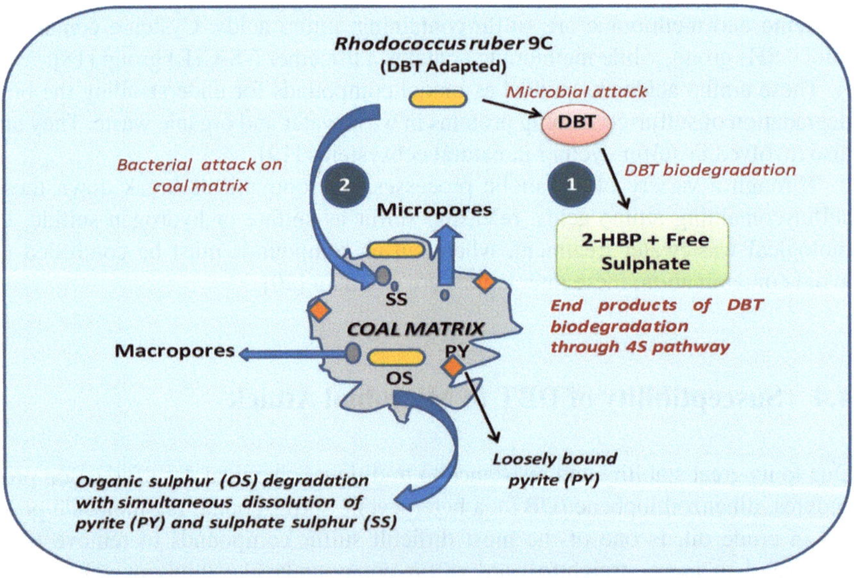

Fig. 4.3 Mechanism of sulfur removal from coal matrix using bacterial attack [26]

4.5 Ring-Destructive Pathways

Ring-destructive pathways refer to the microbial degradation processes that break down the aromatic rings of sulfur-containing compounds, including dibenzothiophene (DBT) and other related molecules. Unlike the selective desulfurization pathways such as the 4S pathway, where the sulfur atom is removed while preserving the hydrocarbon structure, ring-destructive pathways lead to the complete breakdown of the aromatic rings. This often results in the loss of both the sulfur atom and the energy-rich carbon framework, transforming the compound into simpler molecules like CO_2, water, and sulfur byproducts. These pathways are typically less desirable for industrial desulfurization because they eliminate valuable hydrocarbons along with the sulfur, but they are crucial in environmental biodegradation, where complete mineralization of pollutants is needed.

In ring-destructive pathways, microorganisms like certain species of *Pseudomonas* and *Bacillus* can break down DBT and other sulfur-containing compounds by opening the aromatic rings, leading to the formation of non-aromatic intermediates. The first step in this process usually involves the activation of the aromatic ring by enzymes such as dioxygenases, which introduce oxygen into the ring, making it more susceptible to cleavage. Once the ring structure is opened, a series of enzymatic reactions degrade the compound further into simpler molecules like catechols, aliphatic acids, and eventually carbon dioxide (CO_2) and sulfur compounds, such as sulfate or sulfite.

Although ring-destructive pathways are effective for environmental cleanup of sulfur-contaminated sites (bioremediation), they are not ideal for applications such as fuel desulfurization. The destruction of the aromatic rings reduces the fuel's calorific value and alters its properties, making it less efficient for combustion. In industrial processes, the goal is often to retain the hydrocarbon structure while selectively removing sulfur, which is why ring-destructive pathways are more suited to environmental applications where the complete breakdown of pollutants is required. However, these pathways remain an essential part of the sulfur cycle in natural ecosystems, contributing to the detoxification and removal of sulfur pollutants in the environment.

4.6 Sulfur-Specific Pathways

Sulfur-specific pathways, such as the **4S pathway**, are microbial degradation routes designed to remove sulfur from sulfur-containing compounds like dibenzothiophene (DBT) while preserving the hydrocarbon structure of the compound. These pathways are distinct from ring-destructive pathways because they target only the sulfur atom, leaving the carbon skeleton intact, which is especially advantageous in industrial applications such as biodesulfurization of fuels. The most widely studied sulfur-specific pathway is the 4S pathway, used by bacteria like

Rhodococcus and *Gordonia*, which selectively desulfurize DBT into non-toxic and non-polluting products like 2-hydroxybiphenyl and sulfate. This process is beneficial because it removes sulfur without compromising the calorific value or structural integrity of the fuel.

The 4S pathway involves three key steps: oxidation of DBT to dibenzothiophene sulfoxide (DBT-O), further oxidation to dibenzothiophene sulfone (DBT-SO_2), and cleavage of the sulfur-carbon bond to form 2-hydroxybiphenyl (2-HBP) and sulfate. The microbe can use the sulfur from DBT for growth and metabolism thanks to this process, but the carbon backbone is unaltered. The energy-rich components of the fuel are maintained by maintaining the hydrocarbon rings, which makes this technique a very efficient way to desulfurize fossil fuels and lower sulfur dioxide (SO_2) emissions during burning. Because 2-hydroxybiphenyl and sulfate, the end products, are non-toxic and safe for the environment, this method is a desirable choice for producing greener fuel.

Because they provide a biological remedy for the drawbacks of chemical desulfurization techniques like hydrodesulfurization (HDS), which necessitates high temperatures and pressures, the sulfur-specific routes are of great interest. Sulfur-specific pathways are also more sustainable and energy-efficient since they function in milder environments. The availability of appropriate microbial strains and the solubility of sulfur compounds can be limiting considerations, and the rate of sulfur removal through the 4S pathway may be slower than chemical techniques. In order to achieve higher desulfurization rates while preserving the benefits of biological processes for the environment and the economy, ongoing research attempts to improve the efficiency of sulfur-specific pathways through strain improvement, bioreactor optimization, and combining these pathways with other biotechnological or chemical methods.

4.6.1 Anaerobic Sulfur-Specific Pathway

Microbial mechanisms that extract sulfur from sulfur-containing substances in an oxygen-free (anaerobic) environment are referred to as the anaerobic sulfur-specific pathway. Anaerobic sulfur-specific routes use other electron acceptors, such as nitrate, sulfate, or carbon dioxide, to drive sulfur removal in settings lacking oxygen, in contrast to aerobic pathways, such as the 4S process, which depends on oxygen to oxidize and remove sulfur. In anaerobic environments, such as underground oil reserves, sediments, and wastewater treatment systems, where microbes have adapted to obtain energy from sulfur-containing substances without the need for oxygen, these pathways are especially pertinent. The degradation of sulfur compounds in such environments is essential for bioremediation, biogas purification, and the reduction of sulfur pollution.

One of the key microbial groups involved in anaerobic sulfur-specific pathways is **sulfate-reducing bacteria (SRB), such as *Desulfovibrio* and *Desulfobacter*. These microorganisms use sulfate (SO_4^{2-}) as an electron acceptor to reduce sulfur compounds into hydrogen sulfide (H_2S). In the process, sulfur from organic molecules like thiols, thiophenes, or sulfates is converted into H_2S, which can be further processed or removed. SRB play a vital role in anaerobic environments where sulfur compounds are abundant, helping to break down complex organic sulfur compounds while using the sulfur as an energy source. However, H_2S is toxic and corrosive, making its management essential, particularly in industrial settings like biogas production and oil extraction, where hydrogen sulfide buildup can cause significant equipment damage and safety risks.

In anaerobic biodesulfurization, an alternative to sulfate reduction is the use of other anaerobic microorganisms that specifically target sulfur atoms in compounds like dibenzothiophene (DBT). These microorganisms metabolize sulfur-containing compounds by utilizing different electron acceptors, such as nitrate or metal ions, without breaking the carbon backbone of the molecule. This preserves the valuable hydrocarbons while removing sulfur under anaerobic conditions. Research on anaerobic sulfur-specific pathways is still evolving, with the goal of improving efficiency in applications like biofuel production, wastewater treatment, and anaerobic bioreactors. Although these pathways tend to be slower than their aerobic counterparts, they hold promise for sustainable sulfur removal in environments where oxygen is scarce.

4.6.2 Aerobic Sulfur-Specific Pathways

Aerobic sulfur-specific pathways are microbial processes that remove sulfur from sulfur-containing compounds in the presence of oxygen, targeting the sulfur atom while leaving the hydrocarbon structure intact. In the context of biodesulfurization, these pathways are highly important, particularly in sectors like petroleum refining, where the objective is to desulfurize fuels without breaking down their carbon skeletons, which are rich in energy. The 4S pathway, the most well-known aerobic sulfur-specific pathway, is extensively researched due to its capacity to remove sulfur from substances such as dibenzothiophene (DBT), a major sulfur contamination in crude oil. For uses like the production of low-sulfur diesel fuels, where the preservation of hydrocarbons is essential, this method is extremely effective.

The 4S route starts when bacteria such as *Gordonia* or *Rhodococcus erythropolis* oxidize DBT. They do this by using oxygen to initiate a sequence of enzyme processes that turn DBT into DBT-sulfoxide and then DBT-sulfone. The sulfur-carbon bonds in these intermediate molecules are broken by further processes, resulting in the production of 2-hydroxybiphenyl (2-HBP) and the release of sulfur as sulfate (SO_4^{2-}). The pathway is perfect for fuel desulfurization because the compound's carbon skeleton is unaltered during this process. This aerobic mechanism differs from anaerobic sulfur-specific processes, which depend on other electron acceptors like nitrate or sulfate, in that oxygen is a necessary reactant in these oxidative steps.

The aerobic sulfur-specific pathways offer several advantages for industrial applications, particularly in terms of environmental sustainability and energy efficiency.

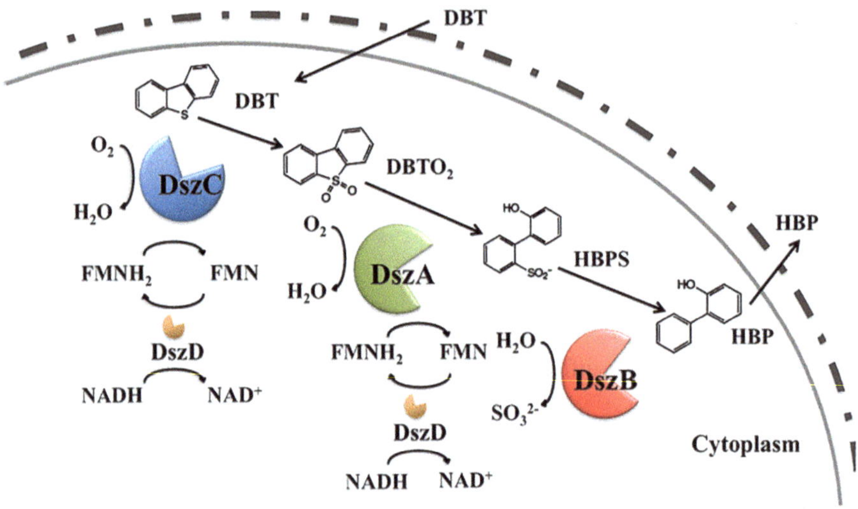

Fig. 4.4 Biodegradation of DBT using 4S pathway [27]

Unlike chemical desulfurization methods like hydrodesulfurization (HDS), which require high temperatures and pressures, aerobic biodesulfurization operates under milder conditions, making it more energy-efficient and environmentally friendly. Additionally, aerobic pathways typically produce non-toxic byproducts, such as 2-hydroxybiphenyl and sulfate, which are easier to manage and dispose of compared to the harmful hydrogen sulfide generated in anaerobic processes. However, aerobic desulfurization can be slower than chemical methods, and ongoing research is focused on enhancing the performance of these microbial pathways through strain improvement, optimizing environmental conditions, and combining biological processes with other desulfurization techniques to improve overall efficiency (Fig. 4.4).

4.6.3 Desulfurizing Microorganisms at a Glance

Desulfurizing microorganisms play a crucial role in the biodesulfurization process, targeting sulfur-containing compounds found in petroleum and other fossil fuels [28]. Key genera include *Rhodococcus*, *Gordonia*, and *Pseudomonas*, which are primarily known for their ability to employ aerobic sulfur-specific pathways such as the 4S pathway. For instance, *Rhodococcus* species can efficiently convert dibenzothiophene (DBT) into less toxic products like 2-hydroxybiphenyl while simultaneously removing sulfur in the form of sulfate. These microorganisms thrive in various environments, making them valuable for industrial applications aimed at reducing sulfur content in fuels, thus minimizing harmful emissions during combustion. *Thiobacillus* and *Desulfovibrio* are two examples of anaerobic desulfurizing bacteria that are crucial to the sulfur cycle. Under anaerobic conditions, these sulfate-reducing bacteria (SRB) convert sulfate to hydrogen sulfide (H_2S) in order to consume organic sulfur compounds. Although this process can produce harmful byproducts like H_2S, it is essential in oxygen-deficient settings like oil reservoirs and deep-sea sediments. *Thiobacillus* and *Desulfovibrio* are two examples of anaerobic desulfurizing bacteria that are crucial to the sulfur cycle. Under anaerobic conditions, these sulfate-reducing bacteria (SRB) convert sulfate to hydrogen sulfide (H_2S) in order to consume organic sulfur compounds. Although this process can produce harmful byproducts like H_2S, it is essential in oxygen-deficient settings like oil reservoirs and deep-sea sediments. In addition to removing sulfur, these microorganisms contribute to the detoxification of pollutants, providing a dual benefit in bioremediation efforts and wastewater treatment.

Desulfurizing microorganisms can break down a broad range of sulfur-containing substances via both aerobic and anaerobic processes because to their varied metabolic capacities. Their uses range from the petroleum sector, where they aid in the production of low-sulfur fuels, to environmental settings, where they aid in the decomposition of sulfur pollutants in polluted areas. Developing long-term solutions to sulfur pollution and the related environmental effects requires ongoing research into how to best optimize these microbes for improved performance and comprehend their ecological responsibilities.

4.6.3.1 *Rhodococcus* spp.

Gram-positive, non-spore-forming bacteria in the genus Rhodococcus are well-known for their ability to break down a wide range of organic substances, including sulfur-containing and aromatic hydrocarbons. These bacteria are especially well-known for desulfurizing dibenzothiophene (DBT) via *the 4S route. In this process, 2-hydroxybiphenyl and sulfate are formed when Rhodococcus oxidizes DBT to DBT-sulfoxide and then to DBT-sulfone. They can also make use of thiophenes and other sulfur compounds. In order to meet environmental rules for low-sulfur fuels, the petroleum sector uses a lot of Rhodococcus strains for the biodesulfurization of fuels. They are great candidates for bioremediation projects because of their resilience and capacity to flourish in a variety of settings.*

4.6.3.2 *Gordonia* spp.

Similar to Rhodococcus, the gram-positive bacterium genus Gordonia is well-known for its ability to break down complex organic contaminants.

Gordonia uses the 4S route for desulfurization, just like Rhodococcus, which removes sulfur from DBT and transforms it into non-toxic compounds. These bacteria have a wide range of metabolic capacities since they can also break down different organic sulfur compounds. In order to produce cleaner fuel and reduce sulfur emissions, Gordonia species are being investigated for use in biodesulfurization processes, particularly in the petrochemical and refining sectors.

4.6.3.3 *Pseudomonas* spp.

Gram-negative bacteria belonging to the varied genus *Pseudomonas* are well-known for their adaptability in metabolism and capacity to flourish in a range of conditions, including soil and water.

Certain species in this genus, such as *Pseudomonas putida*, are able to use substances that include sulfur, such as thiophenes. These bacteria can break down sulfur compounds by ring-destructive pathways and other methods, leading to the full mineralization of the sulfur component, even though they are not usually linked to the 4S pathway.

Because of their capacity to break down a variety of environmental contaminants, particularly sulfur compounds, *Pseudomonas* species are frequently employed in bioremediation and wastewater treatment procedures, lowering toxicity and facilitating cleanup operations..

4.6.3.4 *Desulfovibrio* spp.

Gram-negative, curved rod-shaped sulfate-reducing bacteria (SRB) that flourish in anaerobic conditions are known as Desulfovibrio. The metabolic activities of these bacteria convert sulfate to hydrogen sulfide (H_2S). They are renowned for their capacity to break down sulfated compounds in the absence of oxygen and may also use organic sulfur compounds, such as those present in petroleum.

Desulfovibrio spp. *are primarily studied for their role in biogas production and the anaerobic treatment of wastewater, where they help remove excess sulfate and contribute to the detoxification of sulfur-containing pollutants.*

4.6.3.5 *Thiobacillus* spp.

Obligate chemolithotrophic bacteria of the genus *Thiobacillus* are crucial to the sulfur cycle because they oxidize inorganic sulfur compounds. These bacteria convert hydrogen sulfide or elemental sulfur to sulfate in order to use a variety of sulfur compounds as energy sources. They are essential for the oxidation of sulfur, which allows organic sulfur compounds to mineralize in anaerobic settings. By oxidizing poisonous sulfur species into less hazardous sulfate forms, *Thiobacillus** spp. are utilized in biotechnological applications for bioremediation and the treatment of acid mine drainage, therefore reducing sulfur pollution.

4.6.3.6 *Acidithiobacillus* spp.

A genus of sulfur-oxidizing, acidophilic bacterium called Acidithiobacillus is crucial to bioleaching procedures because it grows well in acidic settings.

These bacteria transform hydrogen sulfide into sulfuric acid and sulfate by oxidizing sulfur and iron to provide energy. They can also help with the bioremediation of contaminated areas by breaking down other organic sulfur compounds.

The mining and mineral processing industries use Acidithiobacillus species, especially in the bioleaching of metals from sulfide ores. By transforming poisonous species into less toxic and more controllable sulfate, they aid in the detoxification of waste that contains sulfur.

4.6.3.7 *Bacillus* spp.

Gram-positive, spore-forming bacteria in the genus *Bacillus* are renowned for their adaptability and capacity to endure in hostile conditions. Sulfur may be extracted from organic substances more easily because of the ability of some *Bacillus* species to break down sulfur-containing amino acids like cysteine and methionine. They frequently use a variety of metabolic processes that aid in the breakdown of sulfur compounds but may not be solely sulfur-specific. Because *Bacillus* species can break down organic contaminants, especially sulfur-containing ones, they are used in biotechnological applications such as soil health enhancement and bioremediation, which helps with environmental cleanup.

4.7 Non-cellular Desulfurizing Biocatalysts

An inventive method of biodesulfurization is represented by non-cellular desulfurizing biocatalysts, which remove sulfur from sulfur-containing materials by using isolated enzymes or enzyme systems obtained from microorganisms rather than entire cells. Compared to conventional microbial methods, this approach has a number of benefits, including improved efficiency, more controlled operation, and better reaction specificity. In these biocatalytic processes, enzymes like sulfite oxidases and dioxygenases play a critical role by enabling the oxidative breaking of sulfur-carbon bonds in substances like thiophenes and dibenzothiophene (DBT). Researchers can increase the pace of desulfurization while reducing byproducts and preserving the integrity of the hydrocarbon structures by utilizing the catalytic capabilities of these enzymes. Simplifying the desulfurization process is one of the main benefits of employing non-cellular biocatalysts. Because enzymes can be immobilized on a variety of substrates, recovery and reuse in many reaction cycles are made easier, potentially resulting in cost savings and improved operational efficiency. The efficiency of enzyme-based systems in eliminating sulfur can also be increased by tailoring them to certain substrates and environmental factors like pH and temperature. The technique is more environmentally friendly and sustainable when non-cellular biocatalysts are used instead of the large amounts of growth media and nutrients needed to cultivate entire cells.

New enzymes with strong desulfurization capabilities are still being sought after and characterized, and developments in protein and genetic engineering should make it easier to create tailored biocatalysts. The stability, selectivity, and activity of these modified enzymes can be improved in situations that are relevant to industry, such high temperatures or the presence of organic solvents. Industries can satisfy regulatory requirements for low-sulfur fuels and reduce emissions and environmental effect by investigating and utilizing the potential of non-cellular desulfurizing biocatalysts to achieve more sustainable and efficient techniques for sulfur removal.

4.8 Dsz Enzymes

Dsz enzymes are essential for the microbial biodesulfurization of sulfur-containing compounds in petroleum and similar products. They are a component of the dibenzothiophene (DBT) desulfurization pathway. Three essential enzymes—DszA, DszB, and DszC—define the Dsz pathway. They function in concert to effectively remove sulfur from DBT and transform it into non-toxic compounds. The first oxidation of DBT to dibenzothiophene sulfoxide (DBT-O) is catalyzed by the flavin-dependent monooxygenase DszA. The sulfur-containing molecule gains oxygen from this reaction, laying the groundwork for further changes. The Dsz enzymes' effectiveness is important because they enable microbes like *Gordonia* and *Rhodococcus* to metabolize DBT and use the sulfur that is produced for growth while maintaining the hydrocarbon backbone. DszB, another enzyme in the process, catalyzes the transformation of DBT-O into dibenzothiophene sulfone (DBT-SO_2) once DszA has finished its job. Because it further oxidizes the intermediate and gets it ready for cleavage, this step is essential. The cleavage of the sulfone by the last enzyme, DszC, results in the formation of 2-hydroxybiphenyl (2-HBP) and sulfate (SO_4^{2-}). Because these bacteria have Dsz enzymes, which selectively target and convert DBT without destroying the hydrocarbons' useful carbon skeleton—a crucial benefit in fuel desulfurization processes—they are effective for biodesulfurization.

The goal of growing Dsz enzyme research is to increase the enzymes' stability and efficiency for use in industry. To produce variations of these enzymes with improved characteristics, like greater substrate selectivity and tolerance to environmental stressors, genetic engineering approaches like directed evolution and protein engineering are being used. Furthermore, the feasibility of employing Dsz enzymes in biocatalytic applications can be further enhanced by the potential for enzyme immobilization approaches. Industries can create more efficient and sustainable ways to remove sulfur by utilizing Dsz enzymes, which will help them comply with strict environmental standards and produce cleaner fuel.

4.9 Sulfur Substrate Specificity

When certain enzymes or microbes can selectively interact with and catalyze reactions involving specific sulfur-containing substances, this is known as sulfur substrate specificity [29, 30]. The goal of biodesulfurization techniques is to efficiently remove sulfur from fuels while reducing the breakdown of valuable hydrocarbon structures, and this specialization is essential. The dibenzothiophene (DBT) desulfurization pathway involves enzymes like DszA, DszB, and DszC that have a high degree of substrate selectivity and mainly target DBT and related molecules like alkyl-DBT and other thiophenic chemicals [5, 31]. Because of their selectivity, the enzymes may effectively catalyze the oxidation and breaking of sulfur bonds without having a substantial impact on the fuel's hydrocarbon structure. The substrate specificity of desulfurizing enzymes is significantly influenced by the molecular structure of sulfur-containing substrates. How well an enzyme can bind and change a particular substrate depends on a number of important factors, including the presence of functional groups, the oxidation state of sulfur, and the steric hindrance surrounding the sulfur atom. Dibenzothiophene derivatives, for instance, may interact preferentially with enzymes because of their aromatic structure and the particular orientation of the sulfur atom [32, 33]. Conversely, because of adverse interactions or an inability to fit the substrate into the enzyme's active site, they might be less active against other sulfur-containing substances, such as aliphatic sulfides or bigger, more complicated thioether complexes [34].

Optimizing biodesulfurization procedures in industrial applications requires an understanding of sulfur substrate specificity. Through techniques like protein engineering and enzyme modification, researchers are looking at ways to expand the spectrum of substrates that desulfurizing enzymes can function on in order to improve their capacity to act on a wider variety of sulfur-containing molecules. This might result in a more thorough and effective removal of sulfur from fuels, resolving the difficulties of adhering to strict environmental standards while optimizing the recovery of priceless hydrocarbon resources. Through customization of these enzymes' substrate specificity, industry can create more sustainable and efficient biotechnological sulfur removal solutions.

4.10 Recent Developments in BDS Mechanism Related to 4S and Kodama Pathway

Our understanding of how microorganisms can efficiently remove sulfur from fossil fuels while maintaining their hydrocarbon structures has improved due to recent advancements in the biodesulfurization (BDS) mechanisms, specifically with reference to the 4S and Kodama pathways. Our understanding of how microorganisms can efficiently remove sulfur from fossil fuels while maintaining their hydrocarbon structures has improved due to recent advancements in the biodesulfurization (BDS) mechanisms, specifically with reference to the 4S and Kodama pathways. Dibenzothiophene (DBT) and similar chemicals can be selectively oxidized into non-toxic products without breaking the carbon backbone using the 4S pathway, which is mostly linked to bacteria like *Rhodococcus* and *Gordonia*. The involvement of important enzymes like DszA, DszB, and DszC has been revealed by recent research aimed at clarifying the enzymatic mechanisms involved in the 4S pathway. The active sites and catalytic processes of these enzymes have been revealed by advances in structural biology techniques like X-ray crystallography and cryo-electron microscopy, which have helped researchers comprehend how to maximize substrate selectivity and reaction efficiency.

Research into the Kodama route, which aids in the desulfurization of DBT, has been accelerated concurrently. Because it uses different intermediates and a different collection of enzymes, the Kodama pathway is not the same as the 4S pathway. Dibenzothiophene dioxygenase is one of the novel enzymes discovered recently that are involved in the first stages of the Kodama pathway. Usually, this process causes sulfur to fully mineralize, with sulfate as a byproduct. Research on the metabolic capacities of bacteria that use the Kodama pathway has brought attention to the possibility of biotechnological uses in bioremediation and waste treatment, especially in settings with high sulfur loading.

Our understanding of the 4S and Kodama pathways has also advanced significantly as a result of the integration of omics technologies, including transcriptomics, metabolomics, and genomics. Researchers have been able to identify sulfur metabolism-related regulatory networks, gene clusters, and metabolic pathways thanks to these methods. Scientists can now evaluate how various enzymes and metabolic pathways interact by using systems biology techniques, which could improve the effectiveness of biodesulfurization procedures. Furthermore, novel approaches of designing microbial strains with enhanced desulfurization capabilities are being made possible by developments in synthetic biology, which will enable customized applications in industrial settings. Scientists can now evaluate how various enzymes and metabolic pathways interact by using systems biology techniques, which could improve the effectiveness of biodesulfurization procedures. Furthermore, novel approaches of designing microbial strains with enhanced desulfurization capabilities are being made possible by developments in synthetic biology, which will enable customized applications in industrial settings. All of these advancements highlight how crucial the 4S and Kodama pathways are to tackling environmental issues related to sulfur emissions and the manufacturing of low-sulfur fuels [35].

4.11 Challenges and Perspectives

Numerous obstacles prevent the biodesulfurization (BDS) processes—especially those involving the 4S and Kodama pathways—from being widely used in industry. One significant problem is that, in comparison to traditional chemical techniques like hydrodesulfurization (HDS), microbial desulfurization has very sluggish reaction rates. The complexity of microbial metabolism and the requirement for particular growth circumstances, such as the availability of nutrients, ideal pH, and temperature ranges, might be blamed for these slower kinetics. Furthermore, the efficacy of biodesulfurization procedures may be limited by the presence of inhibitory chemicals in crude oil, which can negatively impact microbial activity. The creation of more resilient microbial strains or enzymes with improved desulfurization capabilities and resistance to extreme environments are necessary to overcome these obstacles.

The complexity and unpredictability of sulfur-containing chemicals found in crude oil and petroleum products constitute another major obstacle. A multimodal strategy to biodesulfurization is required due to the wide variety of sulfur compounds, including thiophenes, sulfides, and dibenzothiophenes. Different microorganisms may have different substrate specificities, which could result in hazardous intermediate buildup or insufficient desulfurization. Researchers are employing structural and functional investigations to better understand the substrate selectivity of desulfurizing enzymes in order to overcome this problem. Additionally, utilizing multi-enzyme systems or incorporating microbial consortia can improve the breakdown of a wider range of sulfur compounds, increasing the biodesulfurization process' overall efficiency.

Due to the quick advancements in synthetic biology and biotechnology, the future of biodesulfurization technologies is bright. The capabilities of desulfurizing microorganisms are being improved through the use of genetic engineering and metabolic engineering techniques, enabling the development of specialized strains that can efficiently break down a greater range of sulfur compounds in a variety of environmental settings. A deeper comprehension of microbial metabolism and regulatory networks will also be made possible by the use of omics technology, opening the door for creative strategies to maximize biodesulfurization pathways. Combining biotechnological developments with conventional desulfurization techniques may result in more effective and ecologically friendly sulfur removal processes, which would help produce cleaner fuels and lower sulfur emissions as sustainability and environmental regulations tighten.

4.12 Conclusion

Using processes like the 4S and Kodama pathways, biodesulfurization (BDS) is a viable and sustainable method of extracting sulfur from fossil fuels. These microbial methods help produce cleaner fuels while maintaining the integrity of hydrocarbon structures, in addition to meeting strict environmental rules on sulfur emissions. Significant obstacles still exist despite the advancements in our knowledge of the metabolic pathways and enzymatic mechanisms underlying BDS, such as slower reaction rates in comparison to chemical techniques, problems with substrate selectivity, and the existence of inhibitory chemicals in crude oil.

The efficiency and resilience of desulfurizing microorganisms and their enzymes could be greatly increased by biotechnology developments, especially in the areas of genetic and metabolic engineering. Researchers can better understand the intricacies of microbial metabolism and modify strains or enzyme systems to efficiently break down a wider variety of sulfur compounds by employing state-of-the-art methods like omics technology. By combining these biotechnological advancements with conventional desulfurization techniques, sulfur removal procedures may become more sustainable and effective, meeting global sustainability objectives and mitigating the negative environmental effects of sulfur emissions. In the end, overcoming current obstacles will require further research into non-cellular biocatalysts, a better comprehension of the selectivity of sulfur substrates, and the improvement of biodesulfurization routes. The creation of effective biodesulfurization technologies will be essential in forming a more sustainable future as industry work to adopt cleaner production techniques. This will help to reduce the environmental impact of using fossil fuels and provide cleaner energy options.

References

1. Shi W et al (2017) Environmental effect of current desulfurization technology on fly dust emission in China. Renew Sust Energ Rev 72:1–9
2. Srivastava VC (2012) An evaluation of desulfurization technologies for sulfur removal from liquid fuels. RSC Adv 2(3):759–783
3. Alves L et al (2015) Biodesulphurization of fossil fuels: energy, emissions and cost analysis. RSC Adv 5(43):34047–34057
4. Haruna A et al (2022) Sulfur removal technologies from fuel oil for safe and sustainable environment. Fuel 329:125370
5. Sousa JP et al (2020) The bacterial 4S pathway–an economical alternative for crude oil desulphurization that r educes CO2 emissions. Green Chem 22(22):7604–7621
6. Parravicini, F., Characterization of enzymes from desulfurizing bacterial strains. 2015
7. Kilbane JJ (2017) Biodesulfurization: how to make it work? Arab J Sci Eng 42(1):1–9
8. Liu F et al (2020) Effect of microbial nutrients supply on coal bio-desulfurization. J Hazard Mater 384:121324
9. Boniek D et al (2015) Biodesulfurization: a mini review about the immediate search for the future technology. Clean Techn Environ Policy 17:29–37
10. Guchhait S et al (2005) Bio-desulfurization of model organo-sulfur compounds and hydrotreated diesel—experiments and modeling. Chem Eng J 112(1–3):145–151
11. Soleimani M, Bassi A, Margaritis A (2007) Biodesulfurization of refractory organic sulfur compounds in fossil fuels. Biotechnol Adv 25(6):570–596
12. Sadare OO, Obazu F, Daramola MO (2017) Biodesulfurization of petroleum distillates—current status, opportunities and future challenges. Environments 4(4):85
13. Betiha MA et al (2018) Oxidative desulfurization using graphene and its composites for fuel containing thiophene and its derivatives: an update review. Egypt J Pet 27(4):715–730
14. Al-Khazaali WMK, Ataei SA, Khesareh S (2023) Biodesulfurization of fossil fuels: analysis and prospective. F1000Res 8:12
15. Rout PG et al (2022) Study on the reaction mechanism of oxidative microbial desulfurization of organic sulfur-rich coal. Geomicrobiol J 39(3–5):210–218
16. Hou J et al (2023) Sulfur metabolism in Rhodococcus species and their application in desulfurization of fossil fuels. J Appl Microbiol 134(3):plxad048

17. Ofori K (2023) Hydrogen Sulphide. IntechOpen, London
18. Ahmad M et al (2023) State-of-the-art analysis of the fuel desulphurization processes: perspective of CO2 utilization in coal biodesulphurization. Chem Eng J 478:147517
19. Xu J et al (2020) Biodesulfurization of high sulfur coal from Shanxi: optimization of the desulfurization parameters of three kinds of bacteria. Energy Sources Part A Recov Util Environ Effects 42(18):2297–2315
20. Ishfaq HA, Banerjee A, Qamar S (2021) Bio-desulfurization of coal using biotechnological approach, making coal a less harmful fuel. In: Clean Coal technologies: beneficiation, utilization, transport phenomena and prospective. Springer, pp 161–179
21. Nazir MS et al (2021) A review on the methods in diesel desulfurization. Curr Anal Chem 17(6):815–830
22. Faria RG et al (2024) Advanced technologies conciliating desulfurization and Denitrogenation to prepare clean fuels. Catalysts 14(2):137
23. Jatoi AS et al (2021) Overview of bioelectrochemical approaches for sulfur reduction: current and future perspectives. Biomass Convers Biorefinery 13:1–16
24. Kalita M et al (2022) Mechanistic understanding of Gordonia sp. in biodesulfurization of organosulfur compounds. Curr Microbiol 79(3):82
25. Priya A, Dutta K, Daverey A (2022) A comprehensive biotechnological and molecular insight into plastic degradation by microbial community. J Chem Technol Biotechnol 97(2):381–390
26. Mishra S et al (2017) Insights into DBT biodegradation by a native Rhodococcus strain and its Sulphur removal efficacy for two Indian coals and calcined pet coke. Int Biodeterior Biodegradation 120:124–134
27. Li L et al (2019) A kinetic model to optimize and direct the dose ratio of Dsz enzymes in the 4S desulfurization pathway in vitro and in vivo. Biotechnol Lett 41:1333–1341
28. Parravicini F, Brocca S, Lotti M (2016) Evaluation of the conformational stability of recombinant desulfurizing enzymes from a newly isolated Rhodococcus sp. Mol Biotechnol 58:1–11
29. Lakhmir MA et al (2024) Remediation of high sulfur coal using modified alkali-acid hybrid treatment for reduced SOx, low mineral content, and enhanced combustibility: comparative study. Philippine J Sci 153(3):745–755
30. Jatoi AS, Aziz S, Soomro SA (2021) Effect of native microorganism Rhodococcus spp. SL-9 for dibenzothiophene degradation and its application towards clean coal approach. Cleaner Eng Technol 3:100126
31. Jatoi AS, Aziz S, Soomro SA (2021) Experimental and numerical investigation of DBT degradation via Rhodococcus spp.(SL-9) through the use of biological assisted method. Biomass Convers Biorefinery 13:1–9
32. Koyunoğlu C, Karaca H (2023) Microbial desulphurisation of coal: a review. Int J Sustain Energy 42(1):1–24
33. Akram J et al (2024) Genomic analysis and biodesulfurization potential of a new carbon–sulfur bond cleaving Tsukamurella sp. 3OW. Int Microbiol 27:1–16
34. Murarka P, Srivastava P (2020) Biodesulfurization of petroleum wastes. In: Biovalorisation of wastes to renewable chemicals and biofuels. Elsevier, pp 165–187
35. Stylianou M et al (2023) High biodesulfurization efficiency of oil by aerobic (Burkholderia sp. and Serratia sp.) and anaerobic bacteria using various additives. Energy 282:128950

Chapter 5
Enzymology of Biodesulfurization Process

Abstract The enzymology used in the biodesulfurization process offers a sustainable and ecologically friendly alternative to conventional sulfur removal methods from fossil fuels. Specific microbial enzymes enable a series of enzymatic reactions that selectively break down sulfur-containing compounds, such as dibenzothiophene (DBT), while preserving the hydrocarbon structure. DszA, DszB, and DszC are key enzymes in the 4S pathway that catalyze the slow transformation of DBT into non-toxic metabolites such as 2-hydroxybiphenyl (2-HBP). Additional enzyme pathways, such as the Kodama pathway, aid in the complete mineralization of sulfur, demonstrating the versatility of microbial systems in sulfur metabolism.

Recent advances in enzymology have allowed for a deeper knowledge of the structure, function, and catalytic processes of desulfurizing enzymes. This has made it possible to use protein engineering and guided evolution to increase the stability and activity of these enzymes. Furthermore, the combination of omics technology and computational modeling has proved extremely helpful in comprehending the control of enzyme pathways and their interactions within microbial systems. Despite these advancements, issues persist, including as low catalytic efficiency in industrial settings, substrate selectivity, and enzyme inhibition. Innovative biotechnological solutions to these limitations are necessary to optimize the biodesulfurization process and expand it for commercial use. This abstract highlights key enzymatic components and recent advancements in the field of biodesulfurization enzymology to highlight the promise of biodesulfurization as a sustainable alternative to greener energy production.

5.1 Introduction

Biodesulfurization (BDS) is an eco-friendly process that utilizes microorganisms and their enzymatic systems to remove sulfur compounds from fossil fuels [1]. Unlike conventional hydrodesulfurization (HDS), which relies on high temperatures and pressures, BDS operates under mild conditions, making it a sustainable and

A. S. Jatoi, *Biological Removal of Sulphur from Coal*, Green Energy and Technology, https://doi.org/10.1007/978-981-96-8628-5_5

cost-effective alternative [2]. The process exploits the metabolic pathways of sulfur-metabolizing microbes such as *Rhodococcus* and *Pseudomonas* species, which can selectively degrade sulfur-containing compounds without breaking down the carbon backbone of hydrocarbons [3]. This specificity is achieved through a series of enzymatic reactions that target recalcitrant sulfur compounds like dibenzothiophene (DBT) and its derivatives, which are resistant to chemical desulfurization methods [4].

The enzymology of biodesulfurization primarily involves the 4S pathway, a well-studied metabolic route in sulfur metabolism [5]. Key enzymes in this pathway include dibenzothiophene monooxygenase (DszC), dibenzothiophene sulfone monooxygenase (DszA), and desulfinase (DszB) [6, 7]. These enzymes work sequentially to oxidize DBT into 2-hydroxybiphenyl (HBP) and sulfate, effectively removing sulfur without altering the hydrocarbon structure [1]. Enzyme efficiency, stability, and activity under operational conditions are critical to the success of the BDS process. Advances in enzyme engineering, metabolic pathway optimization, and microbial strain development continue to enhance the efficiency of biodesulfurization, paving the way for greener and more sustainable fuel production.

5.2 History and Evolution of Desulfurization Technologies

The development of desulfurization technology began when it was discovered that sulfur emissions had detrimental impacts on both human health and the environment. When sulfur compounds in fossil fuels burn, sulfur dioxide (SO_2) is generated, which is a cause of respiratory illnesses and a precursor to acid rain. Early efforts to reduce the sulfur level in fuels employed simple chemical methods, but these were usually unsuccessful and impractical for widespread usage [1, 8].

The adoption of hydrodesulfurization (HDS) in the mid-1900s marked a significant breakthrough. HDS uses metal-based catalysts (such cobalt or molybdenum) at high pressures and temperatures to extract sulfur as hydrogen sulfide (H2S). Despite its high effectiveness, this method requires a lot of energy and is less effective against refractory sulfur compounds like dibenzothiophene (DBT) and its derivatives. The search for alternative methods resulted from these challenges [9].

In the second part of the twentieth century, biodesulfurization emerged as a promising solution. Microbes may preferentially digest sulfur compounds while preserving the fuel's calorific value, according to preliminary studies. Strains such as *Gordonia alkanivorans* and *Rhodococcus erythropolis* were discovered as a result of research into the separation of sulfur-metabolizing bacteria. These microorganisms use specific enzyme pathways, such as the 4S pathway, to gently and environmentally target sulfur atoms [10].

The transition from laboratory research to commercial applications was fraught with challenges, including low reaction rates and enzyme instability. Advances in genetic engineering, namely, the overexpression of genes associated with desulfurization, significantly improved the survivability of BDS. BDS is increasingly viewed as a complement to conventional methods, particularly in the treatment of refractory sulfur compounds, when paired with developments in bioreactor architecture and process optimization.

A stronger emphasis on sustainability and environmental stewardship is demonstrated by the advancement of desulfurization technologies. By utilizing biotechnological advancements, the field continues to advance toward cleaner energy solutions and supports global efforts to reduce the environmental impact of fossil fuel use.

5.3 Mechanism of Enzymology for the Biodesulfurization

Biodesulfurization (BDS) is a new and environmentally safe method of removing sulfur from fossil fuels by using microorganisms and their enzymatic systems. Compared to normal hydrodesulfurization, which requires significant energy and severe chemical conditions, BDS operates in softer conditions, making it more sustainable and eco-friendly. By focusing on the sulfur atoms in fuel molecules, the process facilitates their selective removal while preserving the hydrocarbon structure. Since it is crucial to process optimization for industrial use, the primary objective of the enzymology of biodesulfurization is to comprehend the specific enzymes involved and their metabolic pathways.

The need for effective desulfurization has grown as a result of stricter environmental regulations designed to reduce sulfur oxide emissions from burning fossil fuels. These pollutants lead to acid rain and environmental degradation. Even though traditional methods like hydrodesulfurization (HDS) have been widely used, their shortcomings—such as high operating costs, energy-intensive requirements, and decreased efficiency for removing refractory sulfur compounds—have prompted research into alternative approaches like BDS. By employing microbial enzymes to specifically target and eliminate sulfur atoms, BDS provides a potential substitute or addition to conventional methods. When it comes to deep desulfurization of fuels that contain complex sulfur compounds, this is particularly true (Fig. 5.1).

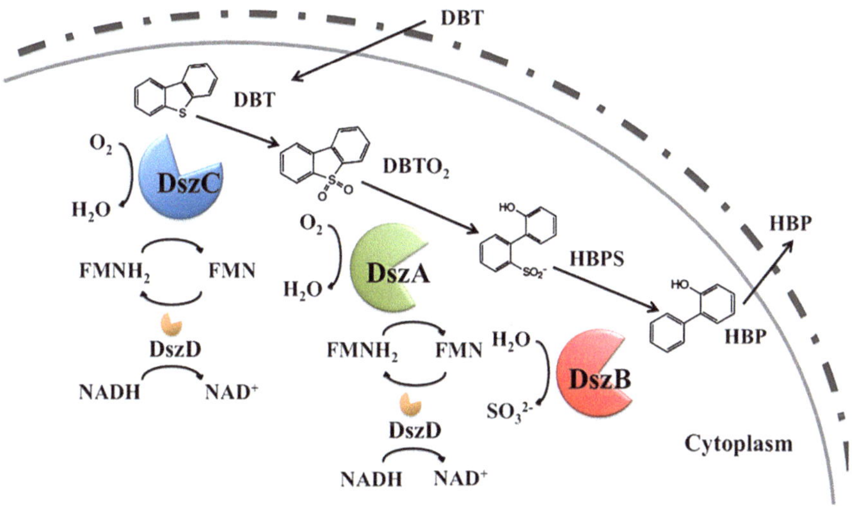

Fig. 5.1 Enzymology of biodesulfurization process [11]

5.4 Genetics of PASHs BDS Pathway

The sulfur oxidation of DBT to DBT-sulfone and other sulfur-containing hydroxylated derivatives as dead-end products is catalyzed by a variety of bacterial ring-hydroxylating dioxygenases and fungal laccases.

5.4.1 *Anaerobic BDS Pathway*

Other anaerobic biodesulfurizing microorganisms (BDM), such as *Desulfomicrobium escambium* and *Desulfovibrio longreachii*, are not known to be able to desulfurize DBT. The advantage of anaerobic BDS of oil is that it releases sulfur as a gas without requiring aeration costs. However, the development of a commercial technique is unlikely to occur anytime soon due to the poor rate and scope of petroleum desulfurization by currently accessible anaerobic cultures and the lack of knowledge regarding the biochemistry and genetics of such bacteria.

5.4.2 *Aerobic BDS Pathway*

5.4.2.1 Kodama Pathway

DBT has been demonstrated to be desulfurized by anaerobic biodesulfurizing microorganisms (BDM) such as *Desulfomicrobium escambium* and *Desulfovibrio longreachii* via a different mechanism than the one biphenyl was involved in. Sulfur is created as a gas and aeration expenses are reduced when oil is BDS anaerobically. However, the poor rate and extent of petroleum desulfurization by currently available anaerobic cultures, as well as the lack of knowledge regarding the biochemistry and genetics of such bacteria, make the development of a commercial strategy improbable in the near future [11].

5.4.2.2 Complete Degradation Pathway

For the degradation of DBT and DBT-sulfoxide (DBTO), respectively, *Arthrobacter* sp. DBTS2 and *Brevibacterium* sp. DO have been identified to use a different type of carbon-carbon cleavage pathway. These two species only get their energy, carbon, and sulfur from DBTO and DBT. The PASHs' desulfurization mechanism, shown in Fig. 5.2, was purportedly somewhat similar to the 4S pathway but was still carbon damaging as they ultimately break down into CO_2, sulfite, and water. It is still unknown which genes and/or enzymes are involved in this pathway. Regretfully, this pathway has not been further studied, and its enzymology and genetics are still unknown. The bacteria that exploit this metabolic pathway may be

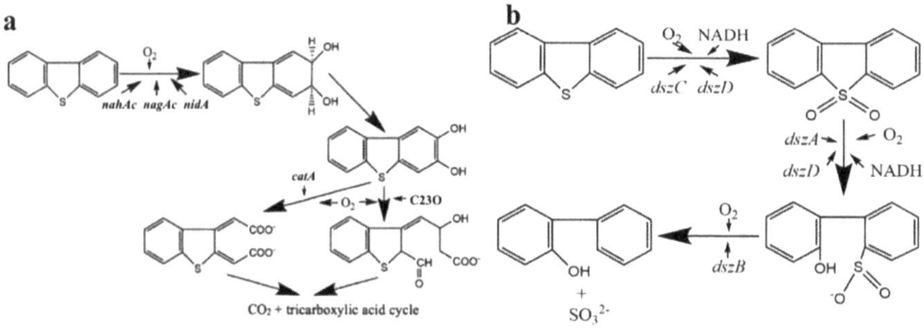

Fig. 5.2 Functional genes for the degradation of DBT (dibenzothiophene) by the Kodama pathway and the 4S pathway. (**a**) Kodama pathway; (**b**) 4S pathway [12]

helpful in creating mixed microbial inocula for the polyaromatic hydrocarbon bioremediation process, despite the fact that it is less significant in terms of BDS technology. Kohler et al. [13] showed that bacteria metabolize hydroxybiphenyls (HBPs) in two main ways. One method employs metacleavage following the hydroxylation of the already hydroxylated aromatic ring by NADH-dependent monooxygenase. The bacteria isolated on 2-hydroxy-and 3-hydroxybiphenyl use this method. The alternative method involves dioxygenating the non-hydroxylated aromatic ring, followed by rearomatization and metacleavage. The FH23 strain of pseudomonas uses this route. A site-specific monooxygenase follows the metabolic pathway by hydroxylating aromatic compounds at the C-3 position when there is a hydroxy group at C-2 and an alkyl or phenyl at C-1, as per Kohler et al. [14], when *Pseudomonas* sp. strain HBP1 breaks down 2-HBP. 2,3-Dihydroxybiphenyl is the process' end product. Meta cells grown on 2,2'-BHBP show the same mechanism, and 2,2′,3-trihydroxybiphenyl is the end product of the monooxygenase attack on 2,2'-BHBP. 2,2′,3-Trihydroxybiphenyl is used as a substrate by monooxygenase activity to produce 2,2′,3,3′-tetrahydroxybiphenyl. As a result, 2,2′,3-trihydroxybiphenyl is hydroxylated at the C-3′ position by the monooxygenase. This demonstrates the extraordinarily broad spectrum of the NADH-dependent 2-HBP monooxygenase enzyme. Fig. 5.3 depicts the suggested process for the metabolism of 2,2'-BHBP. It shows that there are two distinct paths for the metabolism of the first intermediate, 2,2′,3-trihydroxybiphenyl. It can be converted to 2,2′,3,3′-tetrahydroxybiphenyl by the monooxygenase enzyme, which the dioxygenase enzyme can then meta-cleave. The enzyme dioxygenase uses it as a substrate to split the extradiol ring. As a result, both 2,2′,3-tri-and 2,2′,3,3′-tetrahydroxybiphenyl are impacted by the extradiol ring cleavage dioxygenase, producing yellow metacleavage products that are unstable in an aqueous incubation environment. Both compounds could be meta-cleaved by non-enzymatic cyclization. The meta-cleavage product is used as a substrate by hydrolase enzymes to produce 2,3-dihydroxybenzoic acid and salicylic acid, respectively (Fig. 5.2).

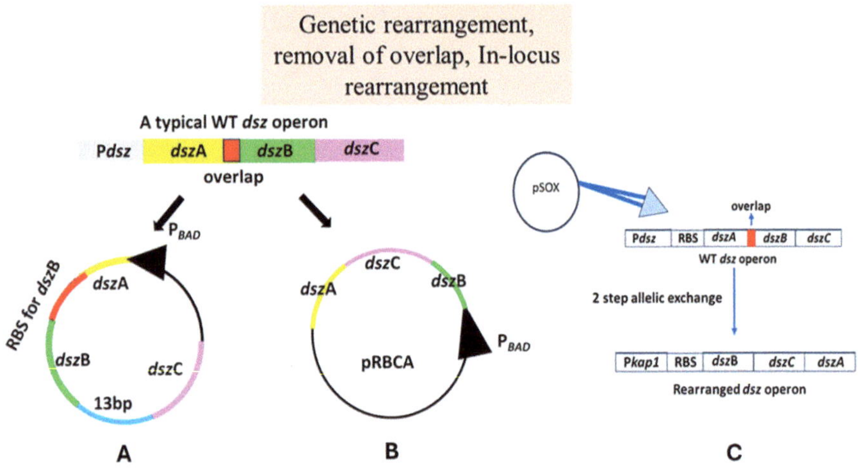

Fig. 5.3 Genetic rearrangement of overlap, In locus rearrangement [3]

5.4.2.3 4S Pathway

In 1985, Isbister and Koblynski discovered that a *Pseudomonas* sp. strain called CB-1 could desulfurize DBT through a sulfur-specific mechanism. Unfortunately, this strain disappeared before the complete identification of the metabolic pathway. The ultimate product was 2-hydroxybiphenyl (2-HBP), with intermediates being DBTO and DBT-sulfone (DBTO2). In 1990, after 40 years of work, Kilbane at the Gas Technology Institute (previously the Institute of Gas Technology) in the United States discovered *Rhodococcus erythropolis* IGTS8. Since then, numerous researchers have identified various bacterial species capable of desulfurizing DBT and its alkylated derivatives via the sulfur-specific process. The "dsZ or 4S pathway" is the name given to the molecular technique of selectively removing sulfur from DBT due to the gradual oxidation of sulfur that takes place across four stages.

5.5 The Desulfurization *dsz* Genes

As seen in Fig. 5.4, DBT is converted into 2-HBP and sulfate in four steps by the dszABC genes, which encode the enzymes. The DszA, DszB, and DszC enzymes are encoded by the 3750 bp long dszABC genes and subsequently translated into polycistronic mRNA. As seen in Fig. 5.3, the dszB and dszC gap is 13 bp, while the dszA termination codon and the dszB initiation codon overlap by 4 bp. dszB seems to express at a lower level than the gene on each side. The DBT monooxygenase (DszC), which is encoded by the dszC gene, catalyzes the conversion of DBT into DBTO and DBTO2. Figures 5.3 and 5.4 illustrates this. The genes dszABC on a

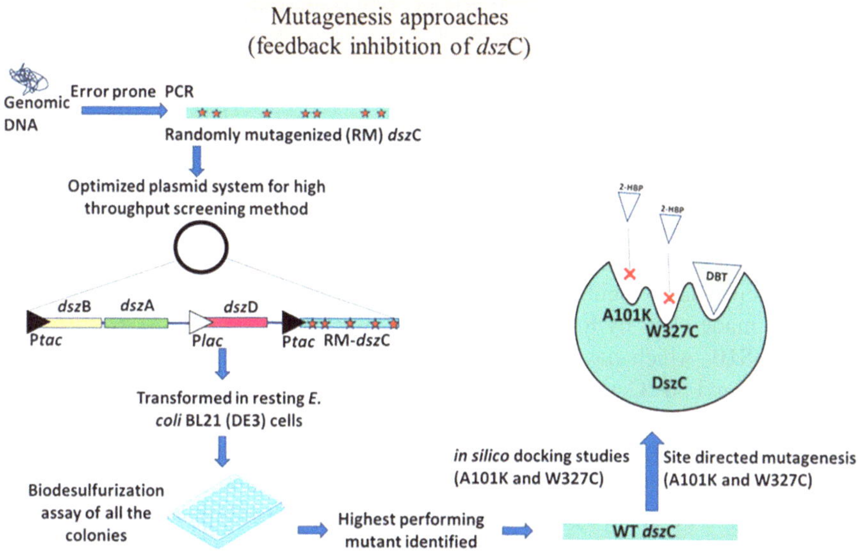

Fig. 5.4 Mutagenesis approaches of the DszC enzyme targeting to reduce the 2-HBP-mediated feedback inhibition [3]

plasmid and dszD on a chromosome were also discovered by the genome sequencing of the desulfurizing bacterium *R. erythropolis* XP. According to Denome et al. [15], the phenotypes of three Dsz genes are grouped together as a 4 kb region on a large plasmid that is organized as one operon and transcribed in a single manner. The 150 kb mega-plasmid that contained the dsz genes grouped in IGTS8 was discovered by Denis-Larose et al. [16]. The genomic DNA of *R. erythropolis* IGTS8 was partially digested by the restriction enzyme TaqI, producing pieces ranging in size from 0.5 to 23 kb. The longer regions were then ligated into the plasmid pRR-6 to create a negative mutant of IGTS8. It was demonstrated that the cloned 6.7 kb region could desulfurize the DBT.

DBT desulfurization, which required a 4 kb BstBI-SnaBI fragment as the minimal DNA sequence, was used to desulfurize IGTS8. Three ORFs were discovered to be present in this DNA fragment: dszA, dszB, and dszC. According to functional investigations for this cluster, the dszC product converts DBT to DBTO2 directly, whereas the dszA and dszB products work together to convert DBTO2 to 2-HBP. Pure DszA, DszB, and DszC enzymes from various bacteria have been used to thoroughly study the dsz pathway.

The analysis of the dszABC genes revealed variations in each gene's termini. Using PCR primers based on the 5 and 3 regions of the *Gordonia amicalis* F.5.25.8 dszABC genes, Kilbane and Robbins [17] were unable to successfully amplify the right genes. The dsz, bds, and tds genes' 5 and 3 termini are not conserved areas. Kilbane and Robbins [17] discovered conserved and unique areas in the three dsz genes, which may be amplified using PCR. The dsz genes from uncharacterized cultures are currently difficult to sufficiently amplify using the PCR method.

The dszABC operon's transcription start point is located 46 nucleotides upstream of the dszA gene's ATG initiation codon. It has been determined that at least three areas influence the expression of dsz: Fig. 7.7 shows three regions that might represent repressor binding site—1) The first section (−263 to 244) has a decrease in Dsz suppression, but its deletion had no effect on the suppression of gene expression; 2) the second region (−146 to −121) may have an activator binding site; and 3) the third region (−98 to −57) may have a repressor binding site. The *Rhodococcus erythropolis* IGTS8 and *Gordonia alkanivorans* RIPI90A dsz operon were found to be quite comparable in their characterization. Bacterial cultures with the same dsz gene sequence have recently been found to display a wide range of Dsz phenotypes. This was made clear by looking at the desulfurization activity of *Mycobacterium phlei* GTIS10, which has the same dszABC gene sequences as *R. erythropolis* IGTS8. Around 50 °C and 30 °C, respectively, were when the cultures showed the greatest desulfurization activity. We described four bacterial cultures that were restricted to utilizing DBT as a source of sulfur. The desulfurization activity, substrate range, and metabolite production of the four cultures were found to vary significantly, even though they shared the identical dsz genes. To better understand the Dsz trait at the intra-species level, significant variations between the genotypic and phenotypic traits of *M. phlei* GTIS10 and *M. phlei* SM120–1 were discovered. The variety of Dsz phenotypes seen in various cultures may indicate how well each bacterial strain or species provides reaction substrates and cofactors. The transfer of substrates and products may also play a role in desulfurization activity, as evidenced by the fact that cell-free lysates of desulfurization cultures can exhibit a larger substrate range than the entire cell culture. Given that the dsz genes are found in many mesophilic bacteria, it is quite likely that they are typically subject to horizontal gene transfer in nature. Desulfurization of thermophilic bacteria has also been demonstrated. According to Ishii et al. [18], *Paenibacillus* sp. A11–2 was the first thermophilic BDM to be discovered; its desulfurization tdsABC genes match those of dszABC. The thermophilic Paenibacillus naphtalenovorans 32O-Y has been found to exhibit additional DszABC/TdsABC identities; nevertheless, these identities are only slightly comparable to the Dsz and Tds proteins from *R. erythropolis* IGTS8 and *Paenibacillus* sp. A11–2 (26–48%, respectively). According to Kirimura et al. [19], the dszABC and tdsABC genes share approximately 58% and 61% of their DNA with the desulfurization genes of thermophilic *Bacillus subtilis* WU-S2B and *Mycobacterium phlei* WU-F1, respectively. Together with the bdsABC genes, overexpression of the protein-producing frm gene was necessary to achieve a high desulfurization efficiency. Over a wide temperature range, the *M. phlei* WU-F1 flavin reductase (Frm) demonstrated strong activity. Additionally, a variety of DBT, benzothiophene (BT), and naphthothiophene compounds can be broken down by *M. phlei* WU-0103.

Both the reducing power (NADH and FMNH2) and the cells' ability to absorb and consume the aromatic compounds had a major impact on the increase in BDS activity. Therefore, despite the fact that neither *Rhodococcus* nor *Mycobacterium* strains have been reported to have DBT transport genes, *Rhodococcus* cells are capable of desulfurizing the extremely hydrophobic chemicals directly from the oil. The fact that DBT is highly selective for the enzymes that desulfurize fossil fuels is one of its drawbacks. To address this issue, significant progress has recently been achieved in creating dsz gene variants that produce novel, hybrid enzymes with improved substrate tolerance. Another issue is that inorganic sulphate, cysteine, or methionine blocks the expression of the dsz genes in their native hosts. In order to address this issue, research in *Rhodococcus erythropolis* IGTS8 showed that the promoter region of the dszABC operon may have a regulatory hairpin structure and that the observed sulfate-mediated repression may be brought on by a putative repressor protein. To bypass sulfate suppression, the tac promoter was utilized to control the dsz genes in multiple strains of *Pseudomonas*, either on a. DBT was effectively desulfurized in *P. aeruginosa* cells thanks to these structures; 95% of 0.2 mM DBT transformed in 24 hours, whereas only 18% of *R. erythropolis* IGTS8 cells did the same.

Both the structure of the molecules being studied and the metabolic activity of the BDSM have an impact on the BDS of DBT derivatives. Since its size and hydrophobicity increase with the alkyl substituent group's carbon number, long-chain alkylated DBT is challenging to desulfurize. However, it has been demonstrated that some bacterial isolates can desulfurize DBTs that are sterically inhibited. Some monooxygenases, like MdsC from *Mycobacterium* sp. G3, have a wide variety of substrates and oxidize various alkyl DBTs to the appropriate sulfone form. The two desulfurizing processes that are present in most bacteria, DBT and BT, are incompatible despite having identical molecular makeups. Only a few bacteria with the dszABC gene can desulfurize both DBT and BT, although it is unclear if Dsz enzymes contribute to this process. Fig. 5.3, which compares the transcriptomes of *Gordonia terrae*, shows that BT is desulfurized by a three-gene operon called bdsABC using a mechanism similar to the 4S process as described (Fig. 5.3).

5.6 Challenges and Enzyme Engineering

The effectiveness of BDS is determined by the stability, specificity, and activity of the enzymes DszC, DszA, and DszB. Weak affinity for certain sulfur-containing molecules, substrate inhibition, and product inhibition are some of the factors that

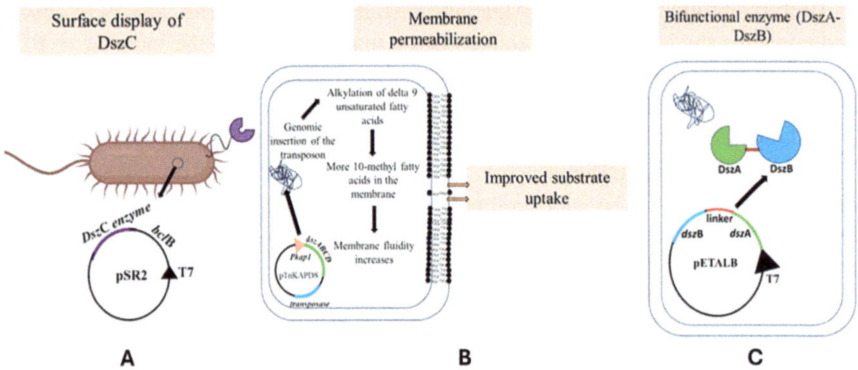

Fig. 5.5 Mechanism of Dsa gene for deulfurization activity

restrict their effectiveness. To get over these obstacles, methods like protein engineering and genetics have been used. Techniques like site-directed mutagenesis and directed evolution have been used to successfully boost catalytic turnover rates, substrate selectivity, and enzyme stability. Modified Dsz enzymes, for instance, are more suited for industrial application because they are more efficient and less prone to inhibition. Researchers have improved the catalytic efficacy of DszC and DszA by introducing specific mutations in the substrate-binding areas or active site. Random mutagenesis in iterative rounds and.

5.7 Operational Conditions and Cofactor Requirements

The presence of inhibitors, temperature, and pH all have a significant impact on the enzymatic activity of the BDS process. Certain pH and temperature ranges are often where enzymes like DszC and DszB show their best activity, necessitating careful bioreactor control. Additionally, cofactors such as flavin adenine dinucleotide (FAD) are necessary for enzyme activity, which emphasizes the importance of maintaining adequate cofactor concentrations in the reaction media. Understanding these factors is necessary to maximize the BDS process' efficiency in eliminating sulfur.

The development of dependable bioreactor systems that can maintain optimal conditions for enzyme activity has been a key focus of research. Immobilization techniques, for example, have been studied to increase the stability and recyclability of enzymes, enabling continuous action in industrial environments. Studies have also examined the effects of various additives and co-substrates on enzyme function in an effort to reduce the impact of the inhibitory compounds present in crude oil.

5.8 Industrial Applications and Metabolic Engineering

For industrial-scale applications, robust microbial hosts that can overexpress the Dsz enzymes in demanding conditions are required. Advances in metabolic engineering have led to the production of recombinant strains with enhanced desulfurization capabilities. For example, overexpression of the dsz operon has significantly boosted sulfur removal rates in **Escherichia coli** and other fast-growing microbial platforms. Additionally, integrating BDS into industrial processes is becoming simpler with the development of bioreactors specifically designed for enzyme stability and activity.

Among the challenges related to the industrial use of BDS are cost-effectiveness, process integration, and scalability. Pilot-scale studies have demonstrated the feasibility of using enhanced bioprocesses and altered microbial strains to desulfurize real fuel samples. These studies highlight the importance of finding a balance between microbial growth, enzyme expression, and substrate availability in order to achieve efficient sulfur removal. Emerging technologies like genome editing and synthetic biology, which enable the creation of tailored microbiological platforms for specific applications, offer new prospects to enhance the performance of BDS systems.

5.9 Global and Environmental Impacts

The application of biodesulfurization will have a major impact on global energy and environmental goals. By reducing the quantity of sulfur in fuels, BDS helps to reduce sulfur oxide emissions, which leads to cleaner air and less environmental degradation. Additionally, the mild operating temperatures of BDS processes reduce energy use and greenhouse gas emissions in comparison to traditional methods, which aligns with the concepts of green chemistry and sustainable development.

Policies and incentives that promote the use of renewable and environmentally friendly technology have raised interest in BDS as a possible replacement for conventional desulfurization methods. To promote BDS research and facilitate its acceptance globally, governments, industry, and academia must collaborate.

5.10 Future Directions

Future studies on biodesulfurization will focus on finding solutions to current problems and exploring new creative possibilities. Emerging methods like high-throughput screening and AI-driven enzyme design are expected to speed up the creation of new desulfurization enzymes. Multidisciplinary methods that combine enzymology, material science, and nanotechnology may enable the creation of complex biocatalysts and bioreactor systems.

Another exciting field of study is the investigation of alternate routes and microbial consortia that can desulfurize a variety of sulfur compounds. Researchers hope to create adaptable and effective BDS systems for a variety of uses by utilizing the metabolic diversity of both naturally occurring and artificially created microorganisms.

5.11 Conclusion and Summary

The enzymology of biodesulfurization is crucial to the development of sustainable fuel purifying techniques. By understanding the molecular foundations of the 4S pathway and enhancing the activity of its key enzymes, researchers are establishing the foundation for efficient and scalable BDS processes. Future advancements in enzyme engineering, process optimization, and microbial host production may displace traditional desulfurization methods, leading to cleaner energy production and environmental protection.

References

1. Jatoi AS, Aziz S, Soomro SA (2021) Effect of native microorganism Rhodococcus spp. SL-9 for dibenzothiophene degradation and its application towards clean coal approach. Cleaner. Eng Technol 3:100126
2. Boniek D, Figueiredo D, dos Santos AFB, de Resende Stoianoff MA (2015) Biodesulfurization: a mini review about the immediate search for the future technology. Clean Techn Environ Policy 17:29–37
3. Bagchi A, Srivastava P (2024) Genetic and metabolic engineering approaches for enhanced biodesulfurization of petroleum fractions. Front Bioeng Biotechnol 12:1482270
4. Jatoi AS, Aziz S, Soomro SA (2021) Experimental and numerical investigation of DBT degradation via Rhodococcus spp.(SL-9) through the use of biological assisted method. Biomass Convers Biorefinery 13:1–9
5. Kalita M, Chutia M, Jha DK, Subrahmanyam G (2022) Mechanistic understanding of Gordonia sp. in biodesulfurization of organosulfur compounds. Curr Microbiol 79(3):82
6. Jatoi AS, Hashmi Z, Anjum A, Bhatti ZA, Siyal SH, Mazari S et al (2021) Overview of bioelectrochemical approaches for sulfur reduction: current and future perspectives. Biomass Convers Biorefinery 13:1–16
7. Jatoi AS, Aziz S, Soomro SA (2024) Bioprocess development and optimization of parameters for enhancing bio-assisted method for sulfur removal from Lakhra coal using Rhodococcus sp SL-9. Int J Coal Prep Util 44(1):37–50
8. Kiran N, Abro R, Abro M, Shah AA, Jatoi AS, Bhutto AW et al (2019) Extractive desulfurization of gasoline using binary solvent of Brønsted-based ionic liquids and non-volatile organic compound. Chem Pap 73:2757–2765
9. Abro R, Kiran N, Ahmed S, Muhammad A, Jatoi AS, Mazari SA et al (2022) Extractive desulfurization of fuel oils using deep eutectic solvents–a comprehensive review. J Environ Chem Eng 10(3):107369

10. Usto MA, Abbasi SA, Maitlo G, Iqbal A, Jatoi AS, Ahmed S et al (2022) Desulfurization of thar lignite by oxidative alkali leaching under pressure. Int J Coal Prep Util 42(11):3430–3450
11. Behnke K, Janssen M (2020) Boundary conditions for traceability in food supply chains using blockchain technology. Int J Inf Manag 52:101969
12. Wang L, Ji G, Huang S (2019) Contribution of the Kodama and 4S pathways to the dibenzothiophene biodegradation in different coastal wetlands under different C/N ratios. J Environ Sci 76:217–226. https://doi.org/10.1016/j.jes.2018.04.029
13. Kohler HP, Kohler-Staub D, Focht DD (1988) Degradation of 2-hydroxybiphenyl and 2, 2'-dihydroxybiphenyl by Pseudomonas sp. strain HBP1. Applied and environmental microbiology 54(11), pp. 2683–2688
14. Kohler HP, Van der Maarel MJ, Kohler-Staub D (1993) Selection of Pseudomonas sp. strain HBP1 Prp for metabolism of 2-propylphenol and elucidation of the degradative pathway. Applied and environmental microbiology 59(3), pp. 860–866
15. Denome SA, Oldfield C, Nash LJ, Young KD (1994) Characterization of the desulfurization genes from Rhodococcus sp. strain IGTS8. Journal of bacteriology 176(21), pp. 6707–6716
16. Denis-Larose C, Labbé D, Bergeron H, Jones AM, Greer CW, Al-Hawari J, Grossman MJ, Sankey BM, and Lau PC (1997) Conservation of plasmid-encoded dibenzothiophene desulfurization genes in several rhodococci. Applied and environmental microbiology 63(7), pp. 2915–2919
17. Kilbane JJ, Robbins J (2007) Characterization of the dszABC genes of Gordonia amicalis F. 5.25. 8 and identification of conserved protein and DNA sequences. Applied microbiology and biotechnology75(4):843–51
18. Ishii T, Yoshida KI, Terai G, Fujita Y, Nakai K (2001) DBTBS: a database of Bacillus subtilis promoters and transcription factors. Nucleic acids research 29(1), pp.278–280
19. Kirimura K, Harada K, Iwasawa H, Tanaka T, Iwasaki Y, Furuya T, Ishii Y, Kino K (2004) Identification and functional analysis of the genes encoding dibenzothiophene-desulfurizing enzymes from thermophilic bacteria. Applied microbiology and biotechnology 65(6), pp. 703–713

Chapter 6
Biodesulfurization Process

Abstract One model system for reducing the sulfur content of petroleum in an environmentally safe manner is the biological desulfurization (biodesulfurization) of dibenzothiophene (DBT) via the 4S pathway. Although the 4S pathway's efficiency is still too low for a commercial oil biodesulfurization process despite a lot of work, it may soon be utilized for commercial processes that produce antibiotics, polythioesters, surfactants, and other chemicals as well as for the detoxification of some chemical warfare agents. Although disulfide-rich proteins are difficult to produce in microbial hosts, proteins with disulfide links are resistant to temperature, pH, and solvents. The 4S pathway can be studied to learn how to produce as many disulfide-rich proteins as possible. By raising the dietary need for sulfur, engineering the operon encoding the 4S pathway to contain higher levels of methionine and cysteine may be able to connect using DBT as the only source of sulfur to increased 4S pathway activity. The development of biocatalysts appropriate for an oil biodesulfurization process may come from this approach, but research on the 4S pathway can also improve our knowledge of microbial physiology to maximize the activity of a multi-step co-factor-requiring pathway and produce highly stable, industrially relevant enzymes for a variety of uses.

6.1 Introduction

The process of producing biocatalysts, which frequently take the place of conventional chemical catalysts, entails growing and creating biological molecules or entire cells that speed up biochemical reactions. These biocatalysts, which can be enzymes, bacteria, fungi, or even plant cells, are all carefully selected for their capacity to carry out specific reactions under gentle working conditions and with high selectivity.. Their uses are numerous and include the food, pharmaceutical, environmental, and biofuel sectors. They may carry out reactions in settings that are frequently safer, more energy-efficient, and less harmful to the environment than conventional chemical processes.

A. S. Jatoi, *Biological Removal of Sulphur from Coal*, Green Energy and Technology, https://doi.org/10.1007/978-981-96-8628-5_6

Finding an organism or enzyme that has the required catalytic activity is usually the first step in the manufacture of biocatalysts. Scientists can modify these catalysts for particular commercial uses by enhancing their stability, activity, or selectivity using methods like genetic engineering and molecular cloning. After production has been perfected, it frequently moves to large-scale fermentation, where the enzyme or microbe is grown under carefully monitored conditions. To optimize yield and efficiency, critical parameters like pH, temperature, and nutrient availability are meticulously controlled. When enzyme manufacturing is involved, the enzymes are then isolated, refined, and prepared for application.

Biotechnology breakthroughs, especially in the areas of protein engineering and recombinant DNA technology, have transformed the manufacture of biocatalysts. These methods make it possible to create new enzymes with specific roles, facilitating reactions that were previously challenging or impossible to accomplish. Additionally, biocatalysts can be immobilized on solid supports to improve their stability and reusability, which increases their industrial applications' cost-effectiveness. As industries work to lessen their environmental effect and transition to greener, more renewable manufacturing techniques, the significance of this sustainable approach to catalysis keeps growing.

6.2 Bioreactors for BDS Studies

In biodesulfurization (BDS) research, bioreactors are essential because they offer a regulated setting for the biological extraction of sulfur from fossil fuels. An environmentally beneficial substitute for traditional sulfur removal methods, which frequently call for severe temperatures and produce hazardous byproducts, is biodesulfurization. In BDS, sulfur-containing compounds in fossil fuels are broken down by specialist bacteria like *Gordonia* or *Rhodococcus*, lowering sulfur levels while maintaining the fuel's calorific value. An essential instrument for optimizing microbial sulfur removal is the bioreactor, which provides optimal conditions for temperature, pH, and oxygen levels.

BDS research frequently employs a variety of bioreactor designs, including fluidized-bed reactors, packed-bed reactors, and stirred-tank reactors. Because they allow for uniform mixing, stirred-tank reactors—which are frequently employed in industrial bioprocessing—are useful for preserving constant conditions throughout the culture. On the other hand, packed-bed and fluidized-bed reactors are frequently utilized for continuous BDS operations, in which fuel constantly runs over immobilized bacteria, and are made to handle enormous volumes with little energy input. Because each form of bioreactor has unique benefits, researchers can choose the best design depending on variables including process scale, desired response rate, and fuel composition.

BDS research in recent years has concentrated on improving bioreactor efficiency through innovations such as improved bioreactor monitoring systems that use sensors for real-time parameter management and co-culture systems, which use various microorganisms to boost desulfurization rates. Sulfur removal can be further optimized by combining bioreactors with pretreatment steps to eliminate additional impurities. Bioreactors will continue to be at the forefront of BDS technology advancements, offering a basis for environmentally friendly desulfurization procedures that may someday take the place of more polluting conventional techniques in industrial settings.

6.3 Separation of Biocatalyst from Reaction Mixture

A crucial stage in bioprocessing is the extraction of a biocatalyst from a reaction mixture, which guarantees product purity and permits the biocatalyst's possible repurposing. This stage is particularly crucial in industrial settings where particular processes are catalyzed by enzymes or microbial cells, such as in the manufacturing of medications, biofuel, or wastewater treatment. Effective separation techniques eliminate waste, cut expenses, and increase process efficiency.

Depending on the type of biocatalyst, the reaction mixture, and the planned downstream processes, there are a number of different methods for separating biocatalysts. In order to separate soluble enzymes from the reaction mixture, techniques such as ultrafiltration and precipitation are frequently employed. Filtration or centrifugation is frequently useful for entire cells or immobilized enzymes since these cells can be divided according to their size and density. Biocatalysts can be immobilized on solid supports to further streamline the separation process and make it easier to retain or reuse the catalyst without requiring a lot of downstream processing.

Recent developments in biocatalyst separation technologies include membrane-based methods that enable continuous separation and the use of magnetic particles for simple recovery of immobilized enzymes or microbial cells by applying a magnetic field. These techniques improve the overall sustainability of bioprocesses by increasing separation efficiency and lowering the loss of biocatalytic activity. Industries can better utilize biocatalysts in a variety of applications by refining separation techniques, which will promote more economical and environmentally friendly production processes.

6.4 Mass Transfer Studies

Understanding and improving processes where substances transition between phases, such as chemical reactions, separation procedures, or biological systems, requires an understanding of mass transfer research. Diffusion and, in some situations, convection propel molecules to move due to concentration differences, which is known as mass transfer. Because mass transfer directly affects reaction rates, yields, and overall process efficiency, industries can improve efficiency in applications like chemical reactors, biological fermentation, environmental treatment systems, and fuel cells by comprehending it.

The rate at which a substance transitions between phases or between places within a single phase is examined in the research of mass transfer. For instance, the mass transfer rate in solid-liquid systems can affect the rate at which a solute dissolves, which has an effect on chemical and pharmaceutical operations. To sustain microbial or cell viability and productivity in gas-liquid systems, such as oxygen transfer in bioreactors, sufficient mass transfer is necessary. Temperature, concentration gradient, pressure, the characteristics of the transferring material, and the system's shape are important variables that impact mass transfer. In order to forecast mass transfer rates under varied circumstances, these parameters are frequently investigated using models and experiments.

Innovations like the use of membranes in separation procedures, microchannel reactors for quick mixing, and immobilization methods that increase diffusion rates are the result of advanced mass transfer investigations. More precise simulations have also been made possible by improved computational modeling, which aids in the prediction of mass transport processes in intricate systems. Optimizing mass transfer continues to be a major area of attention for companies looking to improve process sustainability and efficiency, offering insights that spur the creation of quicker, cleaner, and more effective production techniques.

6.4.1 Oxygen Mass Transfer and Uptake Rates

In bioprocesses like fermentation and bioreactor systems, where aerobic bacteria or cells need oxygen for growth and product creation, oxygen mass transfer and absorption rates are critical characteristics. Because it dissolves poorly in water, oxygen frequently restricts cellular metabolism in liquid cultures when it is not provided in sufficient amounts. Thus, sustaining high productivity and stable bioprocess conditions requires an understanding of and optimization of oxygen transfer from the gas phase to the liquid phase (mass transfer) and subsequently into the cells (uptake).

The pace at which oxygen may transition from the air or gas phase into the liquid phase of the bioreactor is known as the oxygen mass transfer rate, or OTR. The gas-liquid interfacial area, oxygen solubility, agitation, aeration rates, and the gradient in oxygen concentration between the gas and liquid phases are some of the variables that affect this rate. The rate at which the cells absorb oxygen once it has been dissolved in the medium is known as the oxygen uptake rate or OUR. The growth phase, metabolic rate, and cell type all affect OUR. In order to eliminate oxygen constraints that could lower cell development and production, the OTR for effective aerobic bioprocesses must at least equal or above the OUR.

Increasing agitation and aeration rates to improve the gas-liquid interface, introducing oxygen-enriched air, and employing oxygen carriers such hemoglobin-based solutions or perfluorocarbons are some methods to boost oxygen transfer and uptake. High interfacial areas are provided by sophisticated bioreactor designs, including bubble columns or microbubble technologies, to optimize oxygen dissolution. A key component of bioprocess optimization is striking a balance between OTR and OUR. Accurate regulation of these rates promotes high-yield, productive manufacturing in industrial settings, including specialized chemicals, biofuels, and pharmaceuticals.

6.4.2 *Sulfur Substrate Mass Transfer*

A biotechnological method called biodesulfurization (BDS) employs microorganisms, mostly bacteria, to extract sulfur compounds from fuels like gasoline and diesel. Dibenzothiophene (DBT) and other organosulfur compounds are the process' main target because they are difficult for traditional hydrodesulfurization to remove. The efficiency of BDS depends on the efficient mass transfer of sulfur substrates, which affects the microorganisms' bioavailability of sulfur compounds. Diffusion of the substrate through the liquid phase, transfer across cell membranes, and subsequent metabolic conversion are some of the processes in the process.

Both internal and exterior diffusion are involved in the mass transfer of sulfur substrates in BDS. The transfer of sulfur compounds from the bulk liquid to the surface of the microbial cell is referred to as external diffusion. A number of variables, including the sulfur compound's solubility, agitation, and the presence of surfactants that improve solubility, influence this stage. Conversely, internal diffusion entails the substrate moving past the cell membrane and into the cytoplasm, where sulfur removal is carried out by biocatalytic enzymes. During this stage, elements such as transport proteins and membrane permeability are important. Optimizing reactor design, improving mixing, and employing emulsifiers to enhance the surface area for mass transfer can all boost the efficiency of mass transfer.

The limited solubility of many organosulfur compounds in water, which restricts their bioavailability, is one of the primary obstacles in sulfur substrate mass transfer. Techniques like the use of two-phase systems, in which an organic solvent is added to boost solubility, have been investigated in order to overcome this. Internal diffusion can also be improved by genetically modifying microorganisms to express more effective sulfur transport mechanisms. In order to improve mass transfer rates, reactor types including packed-bed and fluidized-bed reactors are also used to enhance contact between bacteria and the sulfur substrate. Overall, optimizing mass transfer is essential for making BDS a viable alternative to conventional desulfurization methods.

6.4.3 The Coupling of BDS and Other Desulfurization Techniques

By combining the benefits of both biological and chemical processes, biodesulfurization (BDS) and traditional desulfurization techniques like hydrodesulfurization (HDS) and oxidative desulfurization (ODS) offer substantial advantages. Dibenzothiophene (DBT) and other refractory sulfur compounds that are challenging to remove using conventional methods are effectively removed by BDS. Its longer reaction times in contrast to HDS, however, may be a drawback. Deeper desulfurization can be achieved while maximizing cost, efficiency, and environmental effect by combining BDS with other procedures.

In hybrid systems, BDS manages the remaining, more complicated sulfur species under gentler, more ecologically friendly circumstances, while HDS is usually employed to remove readily accessible sulfur compounds under high-pressure, high-temperature conditions. The chemical oxidation of sulfur compounds in ODS, on the other hand, increases their water solubility and facilitates their removal in later phases. The complimentary nature of the processes—BDS removes sulfur selectively, while ODS increases the bioavailability of sulfur substrates—makes the combination of BDS and ODS very appealing.

Multiple configurations, such as sequential, parallel, or even simultaneous operation in multiphase reactors, can be used to accomplish process integration. This method limits the production of undesirable byproducts, decreases hydrogen reliance, and lowers the overall energy consumption of desulfurization. Furthermore, advancements in reactor design and process control are facilitating the practical implementation of coupled systems, paving the way for more sustainable fuel desulfurization solutions.

6.5 Improvement of BDS Performance Using Nanotechnology

By using microorganisms to extract sulfur from fossil fuels selectively, biodesulfurization (BDS) is a viable substitute for traditional hydrodesulfurization (HDS) methods. BDS has drawbacks like poor reaction rates and restricted substrate bioavailability, despite its benefits, which include mild operating conditions and selectivity for refractory sulfur compounds like dibenzothiophene (DBT). Nanotechnology provides creative ways to improve microbial activity, mass transfer, and enzyme stability, all of which improve BDS performance. Nanomaterials can close current gaps in BDS procedures and open the door to more effective desulfurization because of their high surface area-to-volume ratio, special chemical characteristics, and adjustable functionality [1–3].

Hydrophobic sulfur compounds, which normally have limited solubility in aqueous systems, can have their bioavailability increased by using nanoparticles [4].

Sulfur compounds can be encapsulated by nanocarriers such liposomes, micelles, and polymer-based nanoparticles, which will make it easier for the molecules to reach microbial cells [4]. Furthermore, immiscible phase dispersion can be greatly improved by nanoemulsions, which improve interaction between sulfur substrates and biocatalysts. Additionally, silica or carbon-based functionalized nanoparticles can serve as supports for immobilizing biocatalysts, increasing the stability and reusability of enzymes [3].

The capacity of magnetic nanoparticles (MNPs) to promote enzyme recovery and reuse has drawn interest [5]. An external magnetic field can be used to easily extract enzymes bound on MNPs from reaction mixtures, which lowers costs and streamlines downstream processing. Likewise, nanomaterials such as carbon nanotubes and graphene oxide improve electron transport mechanisms, increasing the metabolic efficiency of microorganisms in BDS [6]. BDS performance can be further enhanced by the creation of hybrid nanomaterials that combine many capabilities, including as catalytic activity, adsorption, and electron transport [7].

6.5.1 Nanomaterials for Enhancing Biocatalyst Immobilization and Stability

Enhancing the effectiveness and economy of biodesulfurization (BDS) requires the immobilization of biocatalysts. Compared to their free counterparts, immobilized enzymes are more stable, reusable, and resistant to adverse environmental conditions [8]. An excellent platform for immobilizing the enzymes utilized in BDS is provided by nanomaterials, especially nanoparticles (NPs). Because of their small size, nanoparticles can better penetrate and interact with sulfur substrates, and their high surface area can enhance the number of binding sites for enzymes [9].

To support biocatalysts, materials such as carbon-based nanomaterials (graphene, carbon nanotubes), silica nanoparticles, and magnetic nanoparticles (MNPs) are frequently used [10]. Silica nanoparticles, due to their high surface area and tunable porosity, can be used to create porous matrices that enhance enzyme loading and stability [11]. By functionalizing these nanoparticles with particular chemical groups, effective sulfur substrate interaction can be promoted by increasing the binding affinity between the enzyme and the nanoparticle surface [12]. Similar to this, magnetic nanoparticles have the extra benefit of simple recovery through the use of an external magnetic field, which makes it easier to separate the enzyme from the reaction mixture and lowers operational costs and process complexity [13].

Because of their superior conductivity, carbon-based nanomaterials—like graphene oxide and carbon nanotubes—are perfect for boosting electron transport during biodesulfurization. To aid in the breakdown of certain sulfur compounds in BDS, electron transport must be included [14]. The conductivity of graphene oxide and carbon nanotubes enhances the conversion rates of sulfur compounds and increases microbial metabolism. Furthermore, these materials' hydrophobic properties can encourage the binding of hydrophobic sulfur compounds, increasing the microbes' bioavailability. By acting as a barrier against denaturation and aggregation, nanoparticles also help to protect enzymes and keep them active for prolonged periods of time [15]. BDS methods can achieve greater stability, improved performance, and longer operational lives by using nanomaterials for enzyme immobilization. This makes them more cost-effective in industrial applications [16].

6.5.2 Nanoemulsions and Nanocarriers for Enhanced Substrate Bioavailability

The limited bioavailability of sulfur substrates, especially hydrophobic ones like dibenzothiophene (DBT), which has poor solubility in aqueous conditions, is one of the major obstacles in biodesulfurization [16, 17]. Nanotechnology provides a solution to this constraint by improving the dispersion and transport of sulfur substrates to microorganisms through the use of nanoemulsions and nanocarriers. An oil phase, an aqueous phase, and surfactants combine to form stable, submicron-sized emulsions known as nanoemulsions. These emulsions have the ability to greatly improve hydrophobic substances' solubility, making them more accessible for microbial breakdown. Nanoemulsions increase the interaction between the sulfur substrate and microbial cells by spreading the sulfur compounds equally in the aqueous phase and decreasing their size, which results in higher rates of desulfurization.

Sulfur compounds can be encapsulated in nanocarriers such liposomes, micelles, and polymer-based nanoparticles. By solubilizing the sulfur substrates, these carriers can prevent their breakdown and enable their direct distribution to the microbial cells. For instance, sulfur compounds can be encapsulated within the hydrophobic core of liposomes, which are spherical vesicles made of lipid bilayers, enhancing the stability and bioavailability of the substrate. Surfactants can improve the encapsulation of sulfur compounds by lowering the interfacial tension between the aqueous phase and the sulfur phase, which facilitates improved dispersion and microbial cell access.

Additionally, some functional groups that increase the nanocarriers' affinity for sulfur substrates can be incorporated into their design. In order to ensure that the sulfur substrates are efficiently carried to the biocatalysts, functionalization with hydrophobic groups or sulfur-binding moieties improves the interaction between the carrier and the target sulfur compounds. Furthermore, the effectiveness of desulfurization can be increased by using nanoparticles as carriers to assist regulate the substrate's release rate and feed the microbial cells with a steady and regulated supply of sulfur compounds. Incorporating nanoemulsions and nanocarriers into BDS procedures provides a potent means of resolving substrate solubility problems, increasing the effectiveness and viability of biodesulfurization for large-scale uses.

6.5.3 *Hybrid Nanomaterials for Synergistic Effects in BDS*

A promising technique to improve the effectiveness of biodesulfurization (BDS) is the use of hybrid nanomaterials, which blend various nanomaterial types or integrate nanoparticles with biocatalysts. The synergy between diverse nanomaterials can result in higher catalytic activity, greater electron transfer, and better substrate dispersion, making the BDS process faster and more efficient. For instance, hybrid nanomaterials that combine carbon-based substances like graphene oxide or carbon nanotubes with magnetic nanoparticles can have a number of advantages. While carbon-based materials can improve electron transport during microbial sulfur degradation, magnetic nanoparticles (MNPs) can make it easier to recover and reuse biocatalysts, hence boosting overall BDS performance.

Graphene oxide and magnetic nanoparticles are one example of a hybrid material that not only makes it easier to immobilize biocatalysts but also improves electron transport during the desulfurization process. The great surface area and electrical conductivity of graphene oxide make it an excellent substrate for enzyme loading, while magnetic nanoparticles make it simple to recover enzymes with the use of an external magnetic field. By ensuring that the biocatalysts stay stable, active, and recoverable across long cycles, the combination of these materials lowers costs and boosts process efficiency.

Additionally, metal nanoparticles' catalytic qualities and other nanomaterials' structural advantages can be combined in hybrid nanomaterials. For example, BDS can be supplemented by hybrid nanoparticles made of metal oxides or noble metals, such palladium or platinum, embedded in a carbon nanotube matrix to increase catalytic effectiveness in the oxidative desulfurization stage. By facilitating the effective flow of electrons, these hybrid materials increase microbial metabolism and make it possible to break down refractory sulfur compounds, which are generally resistant to conventional desulfurization methods. The development of hybrid nanomaterials that combine several functions can improve BDS in both its chemical and biocatalytic stages, resulting in fuel desulfurization techniques that are more economical and environmentally friendly.

6.5.3.1 Nano-Bioreactor Design and Engineering for BDS Optimization

In order to maximize the efficiency of biodesulfurization (BDS) processes, bioreactor design and optimization are crucial. Reactor design can incorporate nanotechnology to boost microbial activity, increase mass transfer, and enable effective sulfur removal. Several novel approaches to reactor design are provided by nanobiotechnology, which lowers energy consumption, increases the total desulfurization efficiency, and facilitates improved contact between the sulfur substrate and microorganisms.

Using nanomaterials to build bioreactors, such as adding nanoparticles to the reactor's walls or integrating them into a porous matrix, is one important strategy.

For instance, increasing microbial adhesion and guaranteeing that the microbial cells stay active and productive during the desulfurization process can be achieved by using silica- or carbon-based nanomaterials as a support structure for immobilizing microorganisms. These nanoparticles' large surface area and adjustable porosity promote effective microbial growth and enhanced biocatalyst retention, which enhances performance in batch and continuous BDS processes.

Moreover, nano-enhanced bioreactors can facilitate better substrate diffusion. In many BDS systems, poor mass transfer can be a significant bottleneck, limiting the efficiency of sulfur removal. Nanostructured surfaces can enhance the surface area available for substrate interaction, improving diffusion rates and ensuring that sulfur substrates are efficiently transported to the microbial cells. Nanomaterials such as nanowires or nanotubes can be integrated into bioreactors to create a scaffold that promotes better substrate contact and enhances microbial activity.

Additionally, one energy-efficient way to speed up the desulfurization process is to employ nanomaterials with photocatalytic qualities in bioreactors. When exposed to light, photocatalytic nanomaterials, such titanium dioxide (TiO_2) nanoparticles, can produce reactive oxygen species (ROS), which aid in the breakdown of sulfur compounds and promote microbial decomposition. By lowering the energy needs usually connected with conventional desulfurization techniques, this technology can make BDS more ecologically friendly. The BDS process is about to undergo a revolution thanks to the use of nanotechnology into bioreactor design, which will make desulfurization procedures quicker, more effective, and scalable.

6.5.3.2 Environmental and Economic Implications of Nanotechnology in BDS

In addition to improving the process' technical performance, the application of nanotechnology to biodesulfurization (BDS) has major positive effects on the environment and the economy. Conventional desulfurization techniques, such as hydrodesulfurization (HDS), use costly chemicals like hydrogen, generate a lot of trash, and consume a lot of energy. BDS, on the other hand, is a more environmentally friendly substitute that employs microorganisms to break down sulfur compounds in moderate environments. The method is made even more efficient by combining BDS with nanotechnology, which lowers the desulfurization process' operational costs and environmental impact.

The amount of feedstock needed for desulfurization can be decreased by using nanomaterials in BDS to improve the effectiveness of sulfur removal. Nano-enhanced bioreactors, for instance, can increase microbial activity and mass transfer, enabling quicker desulfurization without the need for high pressures or temperatures. When compared to traditional desulfurization techniques, BDS is a more environmentally responsible choice because it uses less energy and produces fewer undesirable byproducts.

By improving the stability and reusability of biocatalysts, the application of nanomaterials can economically lower operating costs. For example, immobilized enzymes supported by nanoparticles can be recovered and utilized repeatedly, which lessens the need for.

Recent Applications of 4S Pathway in Coal Desulfurization.

Future Perspectives.

Conclusion.

6.5.4 Introduction

In recent years, there has been significant attention focused on the biodesulfurization of coal, particularly organic sulfur removal. This interest stems from the environmental issues caused by the combustion of high-sulfur coal, leading to problems like acid rain due to sulfur dioxide emissions. Coal contains sulfur primarily in three forms—pyritic, organic, and sulfate sulfur—with organic sulfur being the most challenging to remove. Its fine dispersion in the coal matrix makes it resistant to conventional chemical desulfurization, resulting in high costs for desulfurizing these stubborn molecules.

To address these challenges, the use of microorganisms in biodesulfurization processes has emerged as an eco-friendly technology and a clean alternative method for efficient sulfur removal, especially targeting organic sulfur. While previous studies have focused on the removal of pyritic and total sulfur, research on organic sulfur removal from coals has been limited. Some investigations have explored the potential of bacteria (such as *Acidithiobacillus ferrooxidans*, *Acidithiobacillus caldus*, *Leptospirillum ferrooxidans*, *Bacillus subtilis*, *Paenibacillus polymyxa*, *Pseudomonas aeruginosa*) and fungi (*Aspergillus genus*) for biodesulfurization purposes, either as flotation bioreagents or through bioleaching processes.

These microorganisms have demonstrated the ability to produce biosurfactants, which are amphiphilic compounds featuring hydrophilic and hydrophobic components. Biosurfactants play a crucial role in various industrial applications, leveraging their properties to reduce surface tensions, enhance solubility, improve detergency power, increase wetting ability, and boost foaming capacity. Notably, some biosurfactants belong to categories like glycolipids, lipopeptides, phospholipids, fatty acids, neutral lipids, and polymeric or particulate compounds. The majority of these biosurfactants are either anionic or neutral, with only a few exhibiting cationic properties characterized by amine groups. The dual capability of utilizing both organic and inorganic compounds as sole carbon and energy sources, along with biosurfactant production, sets the stage for more efficient and environmentally friendly biodesulfurization processes for coals. In this chapter a detailed description of coal biodesulfurization is discussed along with their improvement strategies perspectives and challenges.

6.6 Biocatalyst Production

Since the 4S pathway is a complicated enzyme system, using pure enzyme systems rather than entire cells for a realistic BDS procedure is not an option. As a result, cells that are at rest are thought to be the finest biocatalysts. The following steps are involved in the preparation of desulfurizing resting cells: culturing the chosen strain in an appropriate medium so as to get cells that demonstrate the highest degree of Dsz activity, extracting these active cells, and employing them as the biocatalysts. The production cost of the biocatalyst and its biocatalytic activity must be improved in order for the BDS method to be economically competitive with the deep HDS process. As previously indicated, because Dsz enzymes are sulfate-starvation-induced proteins, the Dsz phenotype is suppressed by readily accessible sulfur compounds, particularly sulfate. Sulfate contamination of the growing medium in this situation is one of the major obstacles to the widespread generation of desulfurizing resting cells. By substituting DBT for sulfate as the sulfur molecule for growth, repression can be prevented. DBT's high cost, poor water solubility, and growth inhibition by 2-HBP have been suggested to make it commercially impracticable to produce biocatalyst in large quantities. In order to replace DBT, several researchers have looked for a suitable sulfur source. The particular desulfurization ability of *R. erythropolis* KA2–5-1 was examined by Yoshikawa et al. (2002) in a fed-batch growth method using 2-aminoethanesulfonic acid as the sole source of sulfur. After 89 h of cultivation, they attained a specific Dsz activity of 111 mmol 2-HBP kg-cell1 h1 and a cell concentration of 20 g l1. Utilizing cheap ammonium sulfate as the only source of sulfur, the same productivity level was attained using another exponential fed-batch culture technique (Hirasawa et al., 2001; Konishi et al., 2005). When 217 mM ethanol was used as the only carbon source, *Mycobacterium* sp. G3's Dsz activity was expressed in media containing MgSO4 at 0.1 mM (Okada et al., 2001); the cultivation period was shorter than when using DBT, and the Dsz activity was higher than those of DBT. Dimethyl sulfoxide (DMSO) has been described as a sulfur substrate that is not a substrate for the DBT desulfurizing enzymes and does not suppress desulfurization activity in *R. erythropolis* IGTS8 (Li et al., 1996). The de-repression of the dsz operon in the absence of more readily accessible sulfur sources (such as sulfide, sulfate, cysteine, and methionine) results in the expression of DBT desulfurizing activity in the presence of DMSO (Li et al., 1996; Oldfield et al., 1997). Many researchers (Matsui et al., 2002; Abbad-Andaloussi et al., 2003a, b; Luo et al., 2003; Bouchez-Natali et al., 2004; Del Olmo et al., 2005; Ma et al., 2003) have used DMSO as the sulfur source for growth.

In *G. alkanivorans* RIPI90A, the 4S route was used to express the Dsz phenotype, while DMSO served as the sulfur source (Mohebali et al., 2008). DMSO-grown resting cells were more active than DBT-grown resting cells (Mohebali et al., 2007). However, resting *Rhodococcus* sp. 1awq cells cultured on DMSO as the only sulfur source demonstrated Dsz activity to be comparable to that attained with DBT (Ma et al., 2006a). Several researchers have described the utilization of a two-stage procedure to try and create the Dsz phenotype in cells cultured on sulfate as the only sulfur source. Their method involved two distinct stages: growth and Dsz-phenotype induction. After achieving a high-cell density culture at the expense of sulfate/DMSO, the Dsz phenotype was generated in the second stage utilizing DBT (Honda et al., 1998; Chang et al., 2001; Ma et al., 2006a; Mohebali et al., 2008). Zn2's impact on *G. alkanivorans* 1B's growth and DBT desulfurization was investigated by Alves et al. (2008b). The specific productivity of 2-HBP achieved by resting cells grown in the presence of Zn2 was 7.6 times more than that of resting cells grown in the absence of Zn2. As a nutritional supply for low-cost DBT desulfurization by *G. alkanivorans* 1B, Alves et al. (2008a) tested the performance of the hydrolyzates generated by enzymatic saccharification of recycled paper sludge in this context. According to Alves and Piaxao (2014), employing sugar beet molasses as a carbon source improved strain B1's ability to desulfurize DBT. According to Martinez et al. (2015), adding co-substrates enhances the yield and rate of DBT biodesulfurization using the 4S route in *Pseudomonas putida* CECT5279 resting cells. Acetic acid as the co-substrate boosted the intracellular levels of NADH and ATP.

6.7 Bioreactors for BDS Studies

The biodesulfurization of oil by sulfate reducing bacteria (SRB) has been studied using two liquideliquid bioreactors, a stirred tank, and a unique electrostatic-dispersion system (Tsouris et al., 1996). In airlift bioreactors, biodesulfurization of diesel has been researched (Nandi, 2010; Irani et al., 2011). Amin (2011) used a vertical rotating immobilized cell reactor with the bacteria *R. erythropolis* in an integrated two-stage process for the biodesulfurization of a model oil. Also investigated was the biodesulfurization of DBT and gas oil in a bioreactor packed with a silica catalytic bed containing immobilized *Rhodococcus rhodochrous* (Alejandro Dinamarca et al., 2014a). *Rhodococcus globerulus* DAQ3 growing cells have been effectively utilized in an aqueous-organic two-layer partitioning continuous process for the biodesulfurization of diesel fuels (Yang et al., 2007). In this procedure, the organic phase and the aqueous phase that contained the cells were maintained as two layers, and the cells were grown in the aqueous phase under steady-state conditions. While sulfur was extracted by the cells from the organosulfur compounds that diffused from the organic layer into the aqueous phase, water-soluble inhibitory chemicals from cell metabolism or cell lysis in the aqueous phase were washed out of the bioreactor (Yang and Marison, 2005). When it comes to maintaining the BDS activity over an extended period of time, this approach has been demonstrated to have a major benefit over batch and fed-batch operations. Systems with microchannel reactors produce a higher yield of products more quickly than ones with bulk reaction. In a BDS process, a microchannel reactor system was employed; this method used a bacterial cell suspension to degrade an alkylated DBT that was not degraded by the control system. The Dsz rate in the oil/water phase was more than nine times higher than in a batch (control) reaction. *Pseudomonas delafieldii* R-8 has been used to biodesulfurize model oil (DBT-containing n-dodecane) and

HDS-treated diesel in a ceramic microsparging aeration system (Lin et al., 2012). For model oil and hydrodesulfurized diesel, microbubble aeration produced activities of 1.3 mg DBT removed g1 h1 and 277 mg sulfur g1 h1, respectively. These numbers were 1.9 and 1.6 times greater than they would have been using a conventional bubble aeration method. Separating the biocatalyst from the reaction mixture (20.3). Another crucial phase in the BDS process is the cell separation after the reaction. Hydrophobic biocatalysts make it difficult for cells to separate from the organic phase; as catalyst concentrations rise, particle-stabilized emulsions occur as a result (Pacheco, 1999; Mohebali et al., 2007). In order to facilitate oil recovery, it has been proposed that the formation of a stable water/oil emulsion could be avoided (Ayala and Vazquez-Duhalt, 2004). Immobilized cells have also been used in a BDS process that has been published (Naito et al., 2001; Lee et al., 2005); this technique allows for the creation of a two-phase system (immobilized cells and oil) for the desulfurization of oil without the production of particle-stabilized emulsions. Although the free dispersion of microorganisms in a fluid reaction volume optimizes mass transmission, it might be challenging to perform the required separation later on, and often the separation technique affects viability. Enzymes and entire cells that have been immobilized as biocatalysts have benefits such as increased stability, recurrent or continuous usage, simple separation from the reaction mixture, and potential catalytic property tuning. Different separation techniques, such as settling tanks, hydrocyclones, and centrifuges, can be assessed for BDS. These processes, however, take a lot of time and money. Immobilized entire cells can be used in the BDS procedure to address the issues (Guobin et al., 2005b; Hou et al., 2005; Lee et al., 2005; Li et al., 2008d; Derikvand and Etemadifar, 2014). According to Hou et al. (2005), immobilized cells had substantially greater stability and longevity than non-immobilized cells. For the desulfurization of light oil, Chang et al. (2000) immobilized *Gordonia* sp. CYKS1 and *Nocardia* sp. CYKS2 DBT-desulfurizing strains in Celite. The immobilized cells could catalyze biodesulfurization repeatedly in the model oil system for more than 900 h with reactivation; recovery of both the biocatalyst and the desulfurized model oil was straightforward. Naito et al. (2001) entrapped *R. erythropolis* KA2–5-1 cells in a photocrosslinkable pre-polymer (ENT4000). For the purpose of enhancing *P. delafieldii* R-8's Dsz activity in oil/water biphasic systems, Li et al. (2008d) investigated its immobilization in calcium alginate beads. By adsorptive immobilization, Alejandro Dinamarca et al. (2014b) immobilized IGTS8 cells on silica, alumina, and Sepiolite. The number of adsorbed cells, metabolic activity, and biodesulfurization have all been investigated in relation to the immobilization of *Pseudomonas stutzeri* TCE3 via adsorption on various inorganic substrates (Alejandro Dinamarca et al., 2010). Additionally, attempts have been attempted to immobilize bacteria, usually in the form of biofilms coating reaction vessels with high surface to volume ratios, preventing the need for further separation (Ansari et al., 2009). Bacterial cells can be embellished with magnetic nanoparticles to combine the benefits of immobilization, simplicity of separation, and microbial lifespan (Shan et al., 2005; Xu et al., 2011). By combining the benefits of magnetic separation and cell immobilization for the BDS process with the hydrophobic *R. erythropolis* LSSE8–1, Li et al. (2009b) have created a straightforward and efficient method. Ammonium oleate was used to modify these nanoparticles such that hydrophilic magnetic fluids could be created. The cells' surface was heavily absorbed by the nanoparticles, which then coated the cells. In comparison to loose cells, coated cells not only had the same Dsz activity but could also be effortlessly pulled apart by magnetic force from the fermentation broth. Magnetic nanoparticles (Fe_3O_4) do not impact the Dsz activity of nanoparticle-coated bacterial cells, according to Bardania et al. (2013). In order to make it easier to separate the cells after the reaction, strain IGST8 cells were coated with magnetic Fe_3O_4 nanoparticles (Ansari et al., 2009). The adorned cells showed a 56% greater DBT desulfurization activity than the nondecorated cells. Model experiments using black lipid membranes showed that the nanoparticles do, in fact, increase membrane permeability, making it easier for reactants and products to enter and exit, respectively. Immobilized in magnetic polyvinyl alcohol beads is *P. delafieldii* (Shan et al., 2003). The primary benefit was the continued high Dsz activity of the magnetically immobilized cells.

6.8 Mass Transfer Studies

6.8.1 Oxygen Mass Transfer and Uptake Rates

Since oxygen is necessary for cell development and Dsz action, all the microorganisms used for BDS are absolutely aerobic. Therefore, oxygen mass transfer from the gas to the liquid phase is crucial not just for development but also for the cells' ability to use dissolved oxygen as a source of energy (BDS). For the same set of experiments conducted in a commercial bioreactor, oxygen uptake rate during growth of *R. erythropolis* IGTS8 has been determined using two experimental methodologies (dynamic and process methods) (Santos et al., 2006b). Models are created to describe oxygen consumption using the experimental values obtained using both methods. In a stirred tank bioreactor, oxygen uptake and mass transfer rates in cultures of *P. putida* CECT5279 capable of desulfurizing DBT and its derivatives were measured under various transport conditions (Gomez et al., 2006); effects of the mass transfer conditions on cell growth and Dsz capacity were noted.

6.8.2 Sulfur Substrate Mass Transfer

One of the important aspects that affects BDS rates is mass transfer of the substrate (Kobayashi et al., 2001). According to Setti et al. (2003), BDS rates are affected by the pace at which DBT is transferred from the oil phase to the water and finally to *Pseudomonas* cells. Biocatalysts in oil-water polyphasic systems must spend money dispersing the oil fraction in the aqueous phase in order to get organic sulfur. It has been noted that the emulsion produced in BDS bioreactors at increasing cell densities is made up of smaller droplets (Pacheco, 1999). The emulsions are produced physically in some bioreactors, such as mechanically mixed reactors and electrospray reactors. Emulsion stabilization can aid in extending the useful life of the generated emulsion in certain circumstances. By inhibiting droplet coalescence in polyphasic systems, specific bacterial cells—often hydrophobic by nature—can also stabilize oil/water emulsions without altering the interfacial tension (Dorobantu et al., 2004; Mohebali et al., 2007). According to Mohebali et al. (2007), *G. alkanivorans* RIPI90A may effectively stabilize water/gas oil emulsions by adhering to the oil-water interface and by preventing droplet coalescence, all while maintaining the environment's surface tension. A stable and long-lasting creamy emulsion can form under these circumstances, and the strain can directly sequester organic sulfur substrates from the hydrophobic phase with little mass transfer restriction. It has been hypothesized that the uptake of the DBT occurs on the surface of the interface because strain IGTS8 is so hydrophobic (Monticello, 2000; Le Borgne and Quintero, 2003). According to other findings, this surface experiences substantial mass transfer constraints when the biocatalyst cells are adherent to it (Maghsoudi et al., 2001). It has been demonstrated that strain CECT5279 was more susceptible to DBT mass transfer limitation in aqueous and biphasic media compared to strain IGTS8 when desulfurizing resting cells with hexadecane as the model oil (Caro et al., 2007). Mass transfer restrictions reduced the yields achieved with strain CECT5279 in biphasic media, according to Caro et al. (2008b). Additionally, because IGTS8 absorbs substances dissolved in oil, mass transfer limits were less significant with IGTS8 than they were with CECT5279. According to the research, *Rhodococcus* hydrophobicity increases its ability to absorb substances dissolved in this phase (Abbad-Andaloussi et al., 2003a; Caro et al., 2007). It has also been demonstrated that strain CECT5279 growth and BDS yields suffered when DBT was used as a sulfur source in the presence of the oil phase (hexadecane) (Caro et al., 2008a). The results of the BDS experiment, which involved growing cells under biphasic circumstances, were simulated using the Haldane model.

6.9 The Coupling of BDS and Other Desulfurization Techniques

Adsorbents such as modified metal oxides, molecular sieves, and activated carbon can be used in desulfurization to remove sulfur compounds under ambient conditions (Li et al., 2008c). The regeneration of adsorbents uses solvent extraction and air oxidation (Hernandez-Maldonado et al., 2004). These two approaches do have certain drawbacks, though. A novel method for creating clean fuels is the combination of biodesulfurization with adsorption desulfurization. Sulfur compounds are initially adsorbed on bioregenerated adsorbents. DBT has been utilized as a model chemical in studies on the bioregeneration of used adsorbents with *P. delafieldii* R-8

(Li et al., 2005a). Fuel desulfurization using a mixed physical-biological process was researched by Li et al. (2006b). Adsorption was used to extract sulfur compounds during this process, and strain R-8 afterward used regeneration of the adsorbent to biodegrade the chemicals. Cells had particles that were comparable to desulfurization adsorbents in size. Therefore, strain R-8 cells supplemented with Fe_3O_4 nanoparticles were employed to separate regenerated adsorbents and cells (Li et al., 2008c). A unique adsorption-bioregeneration system has been built by combining adsorption and BDS processes with the goal of deeply desulfurizing hydrotreated diesel (Li et al., 2009c). By including strain R-8 cells, numerous adsorbents underwent better regeneration. In order to remove sulfur from heavy crude oil (HCO) and light crude oil, Agarwal and Sharma (2010) looked into the usage of BDS, reactive adsorption, oxy-desulfurization, photo-desulfurization, and solvent extraction (LCO). Maximum sulfur removal was achieved by biodesulfurization carried out in anaerobic conditions, followed by oxy-desulfurization, reactive adsorption integration, and removal of 95.21% of sulfur from HCO and 94.30% from LCO.

6.10 Improvement of BDS Performance Using Nanotechnology

Due to its influence on numerous scientific fields, nanotechnology has garnered a lot of attention in recent years. Examining the nanoscale has encouraged the creation and application of new, affordable technologies for catalytic degradation (Zhao et al., 2011). The current state of nanobiocatalysis shows that these methods have a great deal of future promise (Kim et al., 2008). In order to improve BDS rates, Guobin et al. (2005a) integrated g-Al_2O_3 nanoparticles onto *Pseudomonas delafieldii* R-8 biodesulfurizing cells. The goal was to increase the rate at which sulfur compounds from model diesel oil were transferred to the microbial cell. These nanosorbents could remove DBT from oil phase, and they did so at a pace that was much faster than biodesulfurization. Therefore, the DBT transfer rate from the organic phase to the biocatalyst surface was boosted by the nanosorbents, leading to an increase in BDS rates that were around two times greater than those of the original cells. Zhang et al. (2008b) used a variety of commonly used adsorbents, such as g-Al_2O_3, which could selectively remove DBT from the organic phase. The outcomes demonstrated that g-Al_2O_3 has the ability to rapidly adsorb DBT from the oil phase and transfer it to microbial cells, increasing the Dsz rate. The aggregation of g-Al_2O_3 nanoparticles during and after production, however, restricts their use. For their biological uses, well-dispersed and biocompatible g-Al_2O_3 nanoparticle production is essential. Gum Arabic has been used in the synthesis and modification of these nanoparticles to prevent agglomeration in aqueous solutions (Zhang et al., 2007a). Modified g-Al_2O_3 employed in the experiment had an adsorption desulfurization capacity that was 1.12 times more than that of unmodified g-$A1_2O_3$. Fe_3O_4 nanoparticles with various surfaces have been coated on *P. delafieldii* strain R-8 cells (Shan et al., 2005). Ammonium oleate was used to modify the nanoparticles after they had been created. The monodispersed surface-modified nanoparticles in an aqueous solution did not precipitate for more than 18 months. In addition to having the same Dsz activity as free cells, the coated superparamagnetic cells could also be recycled more than five times.

6.11 Future Perspectives

6.11.1 Refinery Challenges

Refineries are struggling to comply with the new USEPA Tier II standards' fuel sulfur specification and the necessary reduction in aromatics (Song and Ma, 2007). Despite being advantageous for the environment, the new environmental laws regulating the sulfur content of diesel and other transportation fuels provide a significant operational and financial burden for the petroleum refining industry. The challenge the refiners are facing is how to provide the growing market demand for ultra-low sulfur diesel. The quality of the crude and diesel feed streams that are made available to the refiners is deteriorating at the same time. Thus, the refiners must create high-quality diesel products like ultra-low sulfur diesel from inferior feedstocks, which is a difficult job (Stanislaus et al., 2010).

6.11.2 Research Needs

The BDS process for diesel desulfurization has been the subject of extensive research during the past 20 years. The results show that the current diesel biodesulfurization process does not work very well to achieve the ultra-low sulfur levels (15 ppm) demanded by new diesel fuel requirements. One of the most difficult tasks for biotechnology is the creation of bioprocesses for the upgrading of vast amounts of fossil fuels. But every study mentioned in the numerous assessments is a big step toward realizing the biotechnological potential for creating a productive BDS procedure. It has not yet been proven that aerobic bacteria can be used in a petroleum biodesulfurization process on an industrial scale. To create a practical approach, we still need to comprehend a lot more facets of the Dsz process. However, due to the low Dsz rate of known bacteria, the commercialization of the BDS method does not appear to be feasible in the near future. The Dsz rate must be increased by roughly 500 times in order to use the biocatalysts that are now accessible. As a result, finding new biodesulfurizers or genetically altering the bacteria that are already available will be necessary for future advancement (Bhatia and Sharma, 2010b). The design of engineered cells with (1) higher specific Dsz activity, (2) broader substrate range,

(3) higher substrate affinity in biphasic reaction systems containing toxic solvents, (4) activity for a long period of time, and (5) higher thermal tolerance are some key research needs for improving biocatalysts for an efficient and commercial BDS process for petroleum and its fractions (Daz and Garca, 2010; Abin-Fuentes et al., 2013). The availability of desulfurizing biocatalyst genome sequences will make it possible to engineer the metabolism (Ma, 2010). Gene regulation-related issues are still mostly unexplored. Contributions from the host cell are crucial in reaching the greater activity. The water/oil volume ratio is one of the most significant technical hurdles in the development of petroleum biotechnological processes; thus it should ideally be reduced to cut down on operational costs for water handling, separation, and disposal (Foght, 2004). The speed at which the sulfur compounds are transported from the oil phase to the membrane of the bacterial cell is another limiting element in the BDS process. Other significant difficulties that demand attention are the efficient separation of the desulfurized oil fraction, the recovery of the biocatalyst, and its return to the bioreactor. To make a BDS process competitive with the catalytic deep HDS methods now utilized in refineries, it has been claimed that more development is necessary in the aforementioned areas (Kilbane, 2006; Mohebali and Ball, 2008). The integration of the BDS process with existing HDS units in the refineries is currently the most appealing solution for the industrial application of the BDS technology in deep desulfurization of diesel. When BDS and catalytic HDS are integrated, the operational costs will drop, and the process economics will improve.

References

1. Karimi E et al (2017) DBT desulfurization by decorating Rhodococcus erythropolis IGTS8 using magnetic Fe_3O_4 nanoparticles in a bioreactor. Eng Life Sci 17(5):528–535
2. Saleh TA (2015) Applying nanotechnology to the desulfurization process in petroleum engineering. IGI global
3. Montalvan-Sorrosa D, de los Cobos-Vasconcelos D, Gonzalez-Sanchez A (2016) Nanotechnology applied to the biodesulfurization of fossil fuels and spent caustic streams. In: Applying nanotechnology to the desulfurization process in petroleum engineering. IGI Global, pp 378–389
4. Deonarine A, Hsu-Kim H (2009) Precipitation of mercuric sulfide nanoparticles in NOM-containing water: implications for the natural environment. Environ Sci Technol 43(7):2368–2373
5. Bilal M et al (2018) Magnetic nanoparticles as versatile carriers for enzymes immobilization: a review. Int J Biol Macromol 120:2530–2544
6. Mariño MA, Fulaz S, Tasic L (2021) Magnetic nanomaterials as biocatalyst carriers for biomass processing: immobilization strategies, reusability, and applications. Magnetochemistry 7(10):133
7. Darwesh OM et al (2020) Enzymes immobilization onto magnetic nanoparticles to improve industrial and environmental applications. In: Methods in enzymology. Elsevier, pp 481–502
8. Cipolatti EP et al (2016) Nanomaterials for biocatalyst immobilization–state of the art and future trends. RSC Adv 6(106):104675–104692
9. Fortes CC et al (2017) Optimization of enzyme immobilization on functionalized magnetic nanoparticles for laccase biocatalytic reactions. Chem Eng Process Process Intensif 117:1–8

10. Razzaghi M et al (2022) Industrial applications of immobilized nano-biocatalysts. Bioprocess Biosyst Eng 45:1–20
11. Khafaga DS et al (2024) Green nanobiocatalysts: enhancing enzyme immobilization for industrial and biomedical applications. PeerJ 12:e17589
12. Bilal M, Nguyen TA, Iqbal HM (2020) Multifunctional carbon nanotubes and their derived nano-constructs for enzyme immobilization–a paradigm shift in biocatalyst design. Coord Chem Rev 422:213475
13. Sohrabi N, Rasouli N, Torkzadeh M (2014) Enhanced stability and catalytic activity of immobilized α-amylase on modified Fe3O4 nanoparticles. Chem Eng J 240:426–433
14. Bilal M et al (2021) Robust nanocarriers to engineer nanobiocatalysts for bioprocessing applications. Adv Colloid Interf Sci 293:102438
15. Zhong L et al (2020) Production and use of immobilized lipases in/on nanomaterials: a review from the waste to biodiesel production. Int J Biol Macromol 152:207–222
16. El-Shishtawy RM et al (2023) Acrylic fabric and nanomaterials to enhance α-amylase-based biocatalytic immobilized systems for industrial food applications. Int J Biol Macromol 233:123539
17. Ahmad R, Sardar M (2015) Enzyme immobilization: an overview on nanoparticles as immobilization matrix. Biochem Anal Biochem 4(2):1

Chapter 7
Reactor for Biodesulfurization of Coal: Overview of Different Process Parameters

Abstract Biodesulfurization is an environmentally friendly and sustainable approach for removing sulfur from coal, significantly reducing sulfur emissions associated with coal combustion. This process relies on specialized microorganisms capable of metabolizing sulfur-containing compounds, thereby converting them into less harmful forms such as sulfate or elemental sulfur. This abstract discusses the design and operational principles of reactors used in biodesulfurization, highlighting various reactor types, including batch, continuous stirred tank reactors (CSTRs), and fluidized bed reactors. Each reactor type offers unique advantages regarding mass transfer, microbial retention, and operational flexibility, influencing the overall efficiency of sulfur removal. Key process parameters such as temperature, pH, oxygen supply, and nutrient availability are critical in optimizing microbial activity and achieving effective desulfurization. Challenges, including the stability of microbial consortia, scaling up from laboratory to industrial applications, and economic competitiveness, must be addressed to enhance the practical implementation of biodesulfurization technologies. Future research directions are outlined, emphasizing the need for innovative reactor designs, integration with renewable energy sources, and the optimization of microbial performance. This chapter covers the importance of reactors in advancing biodesulfurization as a viable solution for cleaner coal utilization and reducing the environmental impact of coal-based energy production.

Keywords Reactor · Biodesulfurization · Microorganisms · Different process parameters

7.1 Introduction

Coal remains a vital energy source globally, especially in industrial sectors such as electricity generation, steel production, and cement manufacturing [1]. However, one of the major environmental concerns associated with coal combustion is the emission of sulfur compounds, particularly sulfur dioxide (SO_2), which contributes

A. S. Jatoi, *Biological Removal of Sulphur from Coal*, Green Energy and Technology, https://doi.org/10.1007/978-981-96-8628-5_7

to acid rain, air pollution, and respiratory diseases [2]. The sulfur in coal exists in both organic and inorganic forms, and conventional chemical or physical methods for desulfurization often prove inefficient, expensive, or environmentally hazardous. As a result, biological methods, particularly biodesulfurization, have emerged as a promising, eco-friendly alternative to mitigate sulfur emissions.

Biodesulfurization is a biotechnological process that leverages the ability of specific microorganisms to remove sulfur compounds from coal [3]. These microbes, primarily bacteria, metabolize sulfur in coal into environmentally benign products. Unlike conventional desulfurization methods that rely on high temperatures, harsh chemicals, and large amounts of energy, biodesulfurization operates under milder conditions, making it more sustainable [4, 5]. This method not only reduces sulfur emissions during coal combustion but also enhances the overall quality of the coal as a fuel source. The integration of biological processes into coal treatment offers the potential to reduce the environmental footprint of coal-fired power plants [6].

At the heart of the biodesulfurization process is the design and operation of the reactor. A reactor provides a controlled environment where microorganisms can thrive and interact with the coal to remove sulfur [7]. Various reactor configurations, such as batch, continuous stirred tank, and fluidized bed reactors, are used to optimize the desulfurization process [8]. The design of the reactor and the choice of microbial species are critical factors that influence the efficiency and scalability of the process. The configuration must account for factors such as mass transfer, coal particle size, and microbial kinetics to ensure effective sulfur removal [9].

To achieve maximum efficiency, it is essential to understand the process parameters that govern the reactor's operation [10]. Key parameters such as temperature, pH, residence time, coal-to-microbe ratio, and nutrient availability significantly affect the microbial activity and sulfur removal rates [11]. Careful control and optimization of these parameters can improve reactor performance, lower operational costs, and make biodesulfurization a competitive option for industrial applications [12]. The development of reactors tailored for biodesulfurization is a step forward in reducing the environmental impact of coal while maintaining its role in global energy production [13].

7.2 Principles of Biodesulfurization

Biodesulfurization is based on the biological ability of certain microorganisms to remove sulfur compounds from coal through metabolic processes [14]. The sulfur present in coal can be categorized into two main types: organic and inorganic sulfur. Both inorganic sulfur, usually in the form of pyrite (FeS_2), and organic sulfur, which is chemically bonded to the carbon structure of coal, contribute to sulfur emissions during coal combustion [15]. Biodesulfurization primarily targets these sulfur compounds, breaking them down into less harmful forms, typically in the form of sulfate or elemental sulfur, which can be easily separated or further processed [16, 17]. The process relies on specialized sulfur-metabolizing microorganisms, most notably chemolithotrophic bacteria such as *Thiobacillus ferrooxidans* and *Desulfovibrio desulfuricans*. These microbes utilize sulfur compounds as an energy source in their metabolic pathways. For instance, *Thiobacillus* species are effective in oxidizing pyrite and transforming sulfur into sulfate, while *Desulfovibrio* species reduce sulfate into hydrogen sulfide in anaerobic environments. These biological reactions occur under relatively mild conditions compared to chemical desulfurization methods, which often require high temperatures and pressures. The microorganisms used in biodesulfurization can metabolize sulfur either aerobically or anaerobically, depending on the type of bacteria and the specific sulfur compounds involved. Aerobic bacteria, such as *Thiobacillus*, oxidize sulfur in the presence of oxygen, leading to the production of sulfate. In contrast, anaerobic bacteria reduce sulfur compounds in oxygen-free environments, producing hydrogen sulfide. The flexibility of using either aerobic or anaerobic systems enables biodesulfurization to be applied in different industrial settings, offering a versatile solution for sulfur removal.

A significant advantage of biodesulfurization is its environmental friendliness. Unlike conventional chemical or physical desulfurization processes, which require large amounts of energy and generate toxic byproducts, biodesulfurization occurs at lower temperatures and produces minimal waste. Additionally, the microorganisms involved can be sustained and cultivated with low operational costs, making the process economically viable for large-scale applications. The eco-friendly and cost-effective nature of biodesulfurization presents a sustainable alternative to conventional sulfur removal technologies, particularly in the context of reducing the sulfur content of coal used in power generation and other industries (Fig. 7.1).

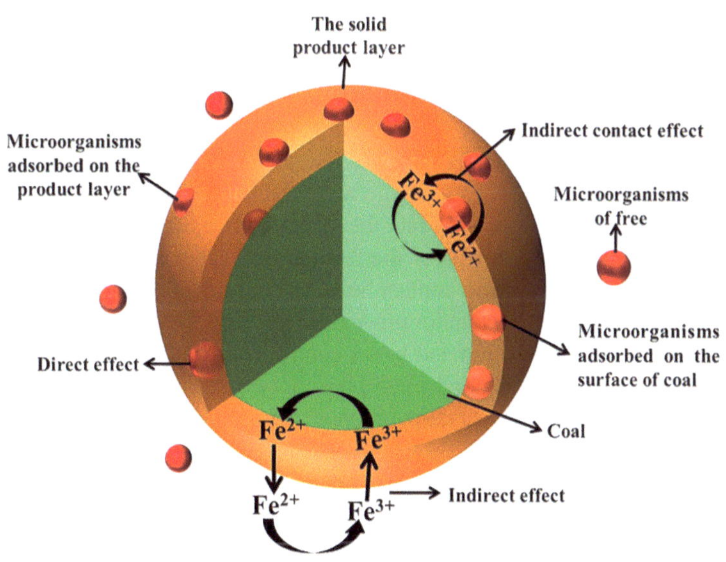

Fig. 7.1 Mechanism of sulfur removal from coal using microorganism [18]

7.3 Reactor Types for Biodesulfurization

The choice of reactor plays a crucial role in the biodesulfurization process as it provides a controlled environment where microorganisms interact with coal to remove sulfur. Various reactor configurations have been developed to optimize the efficiency of this process. Each reactor type is designed to meet specific operational needs, such as maximizing contact between coal particles and microbes, ensuring proper mass transfer, and maintaining suitable conditions for microbial growth. The most commonly used reactor types for biodesulfurization include batch reactors, continuous stirred tank reactors (CSTR), fluidized bed reactors, and packed bed reactors.

7.3.1 Batch Reactors

Batch reactors are one of the simplest designs for biodesulfurization. In this system, coal and microorganisms are placed in the reactor simultaneously, and the reaction occurs over a set period before the entire mixture is discharged. Batch reactors are typically used for small-scale or pilot-scale studies due to their simplicity and ease of operation. However, they can have limitations in terms of scalability and continuous operation. The reaction conditions, such as temperature, pH, and nutrient concentration, need to be carefully controlled to ensure optimal microbial activity. This reactor type is suitable for experimental setups or processes where intermittent operation is acceptable.

7.3.2 Continuous Stirred Tank Reactors (CSTR)

Continuous stirred tank reactors are widely used in industrial biodesulfurization processes due to their ability to maintain a steady-state operation. In a CSTR, coal and microbes are continuously fed into the reactor, while the treated product is continuously removed. The contents of the reactor are kept homogenous by constant stirring, which ensures uniform contact between coal particles and the microorganisms. CSTRs are highly effective for processes requiring constant sulfur removal, as they offer continuous operation and better control over process parameters such as residence time, temperature, and microbial concentration. However, stirring can increase energy consumption and operational costs.

7.3.3 Fluidized Bed Reactors

Fluidized bed reactors are characterized by the suspension of coal particles in the reactor by the upward flow of a liquid or gas, typically air or water. This design allows for excellent mixing and mass transfer between coal and microorganisms, which enhances the biodesulfurization process. The fluidized state also helps in maintaining uniform temperature and microbial distribution throughout the reactor. Fluidized bed reactors are particularly suitable for large-scale industrial applications where continuous operation and high efficiency are required. However, they can be complex to design and operate, requiring careful control of fluid velocity to prevent particle agglomeration or excessive wear on reactor components.

7.3.4 Packed Bed Reactors

In packed bed reactors, coal is packed into a column, and the microbial culture is passed through the bed in either a continuous or recirculating flow. This reactor design offers a high surface area for microbial interaction with the coal and is often used for long-term, stable operations. Packed bed reactors are well suited for processes where the coal particles are small, and the reactor can be operated for extended periods without significant maintenance. However, the potential for clogging and reduced flow rates due to the accumulation of biofilm or sulfur products can be a drawback in this reactor type, requiring periodic cleaning or replacement of the packing material.

Each reactor type offers distinct advantages and limitations depending on the scale of the operation, the characteristics of the coal being treated, and the specific requirements of the biodesulfurization process. The choice of reactor must balance factors such as cost, efficiency, ease of operation, and scalability to achieve the desired level of sulfur removal in a practical and sustainable manner.

7.4 Microbial Species Involved in Desulfurization

Biodesulfurization processes rely on the activity of specialized microorganisms capable of metabolizing sulfur compounds present in coal. These microorganisms, primarily bacteria, play a crucial role in breaking down sulfur-containing compounds into more environmentally benign forms, such as sulfate or elemental sulfur. Different microbial species exhibit distinct metabolic pathways for sulfur oxidation or reduction, which makes them valuable for various biodesulfurization applications. The key microbial species involved in biodesulfurization can be categorized into two main groups: sulfur-oxidizing bacteria (SOB) and sulfate-reducing bacteria (SRB) (Fig. 7.2).

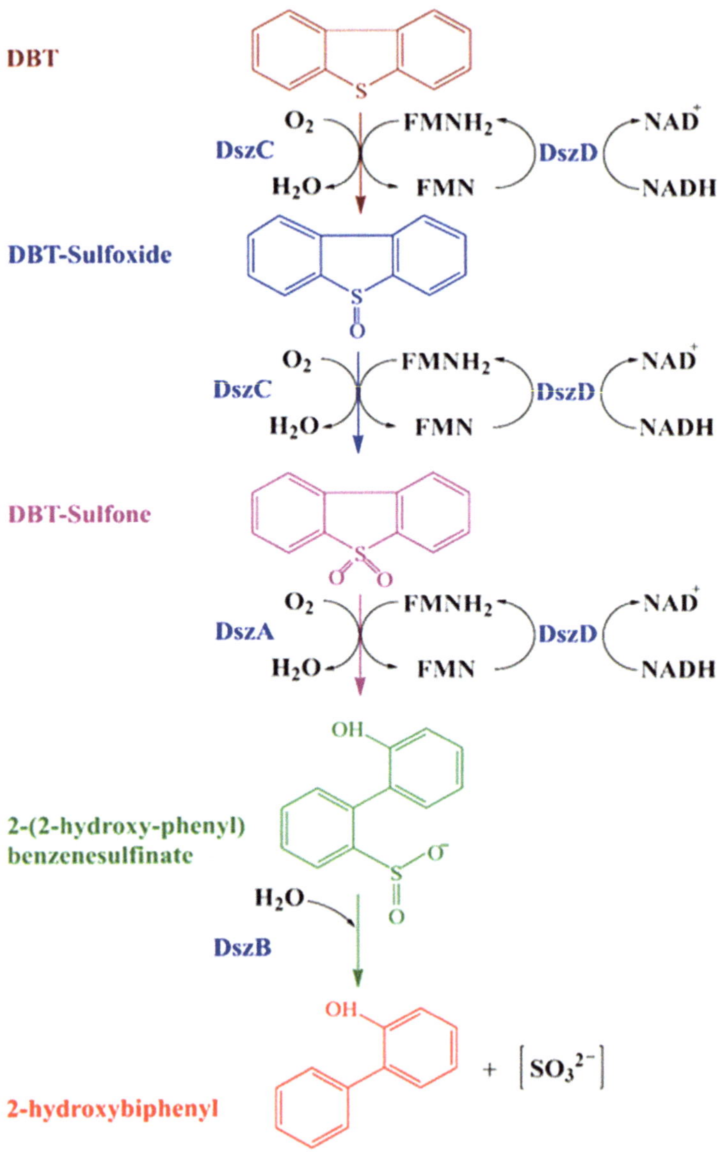

Fig. 7.2 Microbial desulfurization of coal using 4S pathway [19]

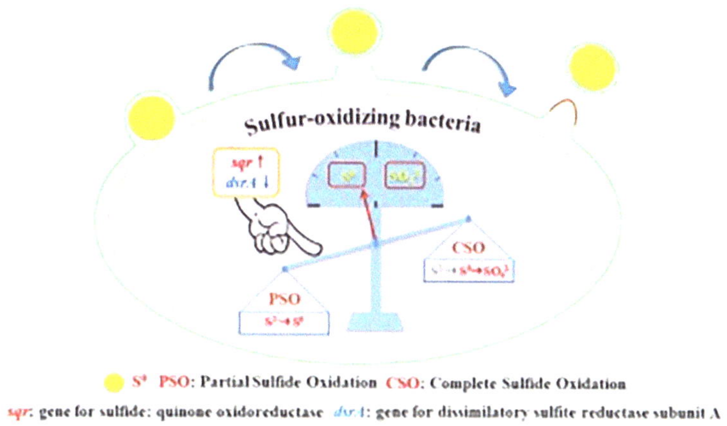

Fig. 7.3 Mechanism of sulfur-oxidizing bacteria [20]

7.4.1 *Sulfur-Oxidizing Bacteria (SOB)*

Sulfur-oxidizing bacteria are aerobic microorganisms that use sulfur compounds, such as pyrite (FeS_2) and organic sulfur, as their energy source. One of the most studied species for biodesulfurization is *Thiobacillus ferrooxidans*, a chemolithotrophic bacterium that thrives in acidic environments. *T. ferrooxidans* oxidizes sulfur compounds into sulfate, making it highly effective at removing inorganic sulfur from coal. Another important SOB is *Acidithiobacillus thiooxidans*, which also oxidizes sulfur compounds and operates efficiently in highly acidic conditions (pH 2–3), typical in biodesulfurization reactors. These bacteria utilize oxygen as their electron acceptor, making them ideal for desulfurization in aerobic reactors.

The biochemical pathway of *T. ferrooxidans* involves the enzymatic oxidation of sulfide or pyrite into sulfate via intermediate sulfur species such as thiosulfate and tetrathionate. This process releases protons, contributing to the acidic environment and further enhancing microbial activity. The efficiency of sulfur oxidation by SOB depends on factors like temperature, pH, and oxygen availability, which must be optimized in reactor systems to ensure continuous sulfur removal (Fig. 7.3).

7.4.2 *Sulfate-Reducing Bacteria (SRB)*

In contrast to sulfur-oxidizing bacteria, sulfate-reducing bacteria are anaerobic microorganisms that thrive in oxygen-free environments. SRB, such as **Desulfovibrio desulfuricans**, reduce sulfate compounds into hydrogen sulfide (H_2S) during their metabolism. This makes them suitable for biodesulfurization processes that target sulfate removal, especially in anaerobic conditions where oxygen is limited.

Desulfovibrio species use organic compounds or hydrogen as electron donors and sulfate as the terminal electron acceptor in their respiration process. SRB play a crucial role in removing both organic and inorganic sulfur from coal, particularly in situations where sulfates are present. They are particularly valuable for deep coal seams or environments where oxygen is not readily available. One challenge with SRB is the production of hydrogen sulfide, which must be further treated or converted to minimize environmental and health risks. Despite this, SRB offer a robust method for sulfur reduction in anaerobic reactors or as part of multi-stage desulfurization processes.

7.4.3 Mixed Microbial Consortia

In many biodesulfurization applications, a consortium of microorganisms is more effective than a single microbial species. A combination of sulfur-oxidizing and sulfate-reducing bacteria can target different forms of sulfur compounds simultaneously. For instance, a reactor may use *Thiobacillus* species for aerobic oxidation of pyritic sulfur and *Desulfovibrio* species for anaerobic reduction of sulfate. These consortia work synergistically to enhance overall sulfur removal efficiency. Mixed microbial systems also increase the versatility of biodesulfurization processes, allowing them to be adapted to various reactor conditions and coal types.

Other microbial species, such as *Pseudomonas putida* and *Rhodococcus erythropolis*, have shown potential in the desulfurization of organic sulfur compounds through enzymatic cleavage of sulfur-carbon bonds (in organic sulfur). These bacteria possess specific enzymes that target thiophenic and sulfidic compounds commonly found in coal.

7.4.4 Factors Influencing Microbial Activity

The activity and efficiency of these microbial species depend on several factors, including temperature, pH, oxygen availability, and nutrient concentration. For example, sulfur-oxidizing bacteria typically thrive at temperatures between 25 °C and 35 °C, while sulfate-reducing bacteria are more active in environments with a neutral to slightly alkaline pH. Nutrient availability, particularly nitrogen and phosphorus, is also critical for microbial growth and sulfur metabolism. Reactor design must take these factors into account to create optimal conditions for microbial activity and sulfur removal.

The microbial species involved in biodesulfurization are key to the success of the process. By selecting the appropriate organisms and creating favorable conditions in the reactor, it is possible to achieve efficient and sustainable sulfur removal from coal. The use of both SOB and SRB, either individually or in combination, allows for a versatile approach to desulfurization, tailored to the specific needs of the application.

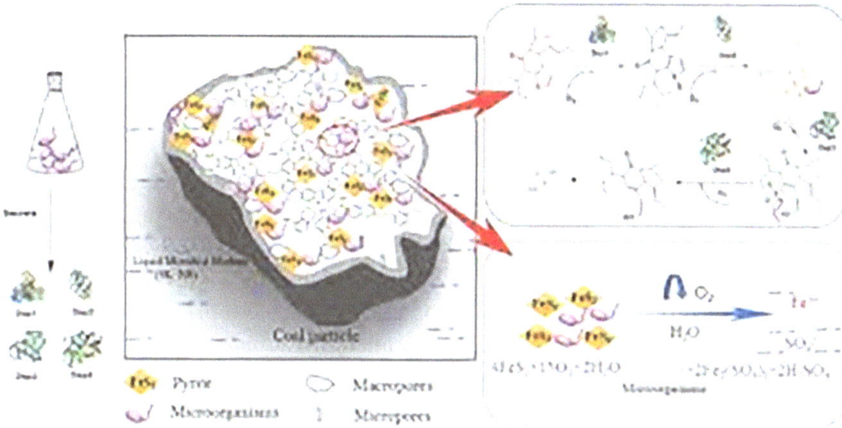

7.5 Process Parameters in Biodesulfurization

Effective biodesulfurization relies heavily on controlling various process parameters within the reactor to optimize microbial activity and ensure efficient sulfur removal from coal. These parameters include environmental conditions such as temperature, pH, oxygen levels, and other operational factors like residence time, coal particle size, and nutrient availability. Proper management of these variables can significantly impact the overall performance of the desulfurization process, influencing both the rate of sulfur removal and the cost-effectiveness of the system.

7.5.1 *Temperature*

Temperature is a critical factor influencing the metabolic activity of microorganisms involved in biodesulfurization. Most sulfur-oxidizing and sulfate-reducing bacteria function optimally within a specific temperature range. For example, **Thiobacillus ferrooxidans**, a key sulfur-oxidizing bacterium, thrives in a range of 25 °C to 35 °C, while sulfate-reducing bacteria like **Desulfovibrio desulfuricans** prefer moderate temperatures between 30 °C and 40 °C. Operating outside these temperature ranges can slow microbial metabolism, reducing the rate of sulfur removal. Furthermore, temperature fluctuations can negatively affect the stability of microbial consortia and the efficiency of enzymatic processes involved in sulfur oxidation and reduction.

7.5.2 pH

The pH of the reactor environment has a profound effect on microbial growth and sulfur metabolism. Sulfur-oxidizing bacteria, such as **Thiobacillus** species, perform best in acidic conditions, often with a pH range between 2 and 4. This acidic environment enhances the solubilization of sulfur compounds like pyrite, allowing microbes easier access to the sulfur substrate. On the other hand, sulfate-reducing bacteria prefer neutral to slightly alkaline pH levels (around 6.5–8), as extreme acidity can inhibit their enzymatic activity. In reactors where both sulfur oxidation and sulfate reduction are taking place, maintaining a balanced pH is essential for the coexistence and efficiency of these different microbial groups.

7.5.3 Residence Time

Residence time refers to the duration coal particles and microbes remain in the reactor, which affects the extent of sulfur removal. In biodesulfurization, longer residence times generally lead to more complete sulfur removal, as the microorganisms have more time to interact with the coal and metabolize sulfur compounds. However, excessively long residence times may lead to diminishing returns, increased operational costs, and inefficiencies in large-scale systems. Residence time is particularly important in continuous processes like those using CSTRs or fluidized bed reactors, where careful control is necessary to optimize the balance between sulfur removal efficiency and throughput.

7.5.4 Coal Particle Size

The size of coal particles introduced into the reactor impacts the surface area available for microbial action. Smaller coal particles provide a larger surface area relative to volume, allowing microorganisms to access sulfur compounds more easily. This enhances the biodesulfurization rate by improving the efficiency of sulfur metabolism. However, extremely fine particles can cause operational issues such as clogging in packed bed reactors or excessive wear in fluidized bed systems. Additionally, smaller particles may require higher energy input for grinding, increasing the overall cost of the process. Finding the optimal particle size is key to balancing sulfur removal efficiency and operational feasibility.

7.5.5 *Coal-to-Microbe Ratio*

The coal-to-microbe ratio is another crucial parameter that dictates the availability of sulfur compounds relative to the microbial population. A high coal-to-microbe ratio can overwhelm the microbial community, resulting in incomplete desulfurization due to insufficient microbial activity to process the sulfur load. Conversely, a low coal-to-microbe ratio can lead to underutilization of reactor capacity and inefficiencies in processing. The optimal ratio ensures that the microbial population has enough coal substrate to metabolize without being overburdened, maximizing sulfur removal efficiency while maintaining a stable microbial environment.

7.5.6 *Oxygen Supply*

Oxygen availability is essential for the activity of sulfur-oxidizing bacteria, which rely on oxygen as a terminal electron acceptor in the process of sulfur oxidation. In aerobic biodesulfurization systems, maintaining adequate oxygen levels is vital to ensure continuous microbial growth and activity. Insufficient oxygen can limit the sulfur oxidation process, leading to lower sulfur removal efficiency. Oxygen is typically supplied via air sparging or forced aeration in aerobic reactors. In contrast, sulfate-reducing bacteria operate under anaerobic conditions, where oxygen presence must be minimized to prevent inhibition of their sulfur-reducing pathways. Reactor designs that incorporate both aerobic and anaerobic zones must carefully manage oxygen levels to accommodate the needs of different microbial communities.

7.5.7 *Nutrient Availability*

Microbial growth and metabolic activity in biodesulfurization reactors depend not only on sulfur compounds but also on the availability of essential nutrients such as nitrogen, phosphorus, and trace elements like iron and magnesium. These nutrients support microbial cell function, enzyme activity, and replication. A lack of nutrients can slow microbial growth and reduce the efficiency of sulfur removal. Typically, nutrient supplements are added to the reactor in the form of mineral salts or organic compounds to ensure sustained microbial activity. However, excessive nutrient concentrations can lead to undesirable side effects, such as biofouling or overgrowth of unwanted microbial species.

7.5.8 Agitation and Mixing

Proper mixing is important in biodesulfurization reactors to ensure even distribution of coal particles, microorganisms, nutrients, and oxygen (in aerobic systems). In stirred tank reactors (CSTRs), agitation keeps the contents homogeneous, enhancing the interaction between microbes and coal. In fluidized bed reactors, the upward flow of gas or liquid ensures uniform contact and prevents particle settling. In packed bed reactors, ensuring sufficient flow of the microbial culture through the packed column is essential to prevent clogging and ensure efficient sulfur removal.

By optimizing these process parameters, biodesulfurization reactors can achieve high sulfur removal rates while minimizing energy consumption, operational costs, and environmental impact. Careful monitoring and control of temperature, pH, residence time, oxygen supply, and other variables are essential for successful biodesulfurization operations, particularly in large-scale industrial applications.

7.6 Environmental Impacts

Biodesulfurization, as an eco-friendly alternative to conventional sulfur removal methods, has notable environmental benefits and impacts. Unlike traditional chemical or physical processes, which often generate hazardous byproducts and require substantial energy, biodesulfurization harnesses the metabolic abilities of microorganisms to remove sulfur from coal under mild conditions. Despite its potential, the process also presents certain challenges that need to be managed to maximize its environmental sustainability.

7.6.1 Reduction in Sulfur Emissions

One of the most significant environmental benefits of biodesulfurization is its ability to substantially reduce sulfur emissions from coal combustion. Coal is a major source of sulfur dioxide (SO_2), which contributes to air pollution, acid rain, and respiratory diseases. By removing sulfur compounds from coal before it is burned, biodesulfurization prevents the release of SO_2 into the atmosphere, thereby reducing acid rain formation and improving air quality. This reduction in sulfur emissions directly contributes to compliance with environmental regulations such as the Clean Air Act and international agreements aimed at lowering greenhouse gas emissions and protecting ecosystems.

7.6.2 Lower Energy Consumption

Compared to traditional desulfurization methods like chemical leaching, physical separation, or flue gas desulfurization (FGD), biodesulfurization operates at lower temperatures and pressures. This significantly reduces the energy required for the process, resulting in a smaller carbon footprint. Since microorganisms metabolize sulfur at ambient or slightly elevated temperatures, there is no need for the energy-intensive processes typically associated with chemical desulfurization, such as heating and pressurizing reactors. This reduction in energy consumption makes biodesulfurization a more sustainable option, particularly when implemented on an industrial scale.

7.6.3 Minimal Chemical Use and Byproduct Generation

Conventional desulfurization methods often involve harsh chemicals that can create toxic byproducts or secondary pollution. For example, chemical desulfurization may use solvents, acids, or bases that result in the generation of hazardous waste requiring further treatment. In contrast, biodesulfurization produces minimal byproducts, and those it does generate, such as sulfate or elemental sulfur, are generally non-toxic and easier to manage. Elemental sulfur, for instance, can be recovered and used in agricultural fertilizers or industrial applications. Additionally, the biological nature of the process means that chemical inputs are minimized, leading to fewer concerns about waste management and potential contamination of surrounding environments.

7.6.4 Wastewater Treatment Challenges

Despite its advantages, biodesulfurization can produce wastewater containing sulfur compounds, such as sulfates or hydrogen sulfide (H_2S), depending on the microbial pathways involved. These sulfurous compounds, especially H_2S, can be harmful to aquatic ecosystems if not properly treated before discharge. Therefore, biodesulfurization processes need to be integrated with effective wastewater treatment systems to neutralize or convert harmful byproducts into non-toxic forms. Hydrogen sulfide, for example, can be oxidized to elemental sulfur or converted into sulfate, which poses less of an environmental risk.

7.6.5 Biodiversity Considerations and Biofilm Formation

The microorganisms used in biodesulfurization, such as **Thiobacillus ferrooxidans** or **Desulfovibrio desulfuricans**, must be carefully managed to avoid unintended environmental consequences. For instance, the release of non-native or genetically modified microorganisms into the environment could disrupt local ecosystems and biodiversity. Additionally, excessive microbial growth in the reactor can lead to biofilm formation, which can reduce reactor efficiency and necessitate more frequent cleaning or maintenance. Biofilm buildup may also contribute to clogging in packed bed reactors or uneven flow distribution in fluidized bed systems, indirectly impacting the environmental sustainability of the process by increasing operational complexity and maintenance requirements.

7.6.6 Carbon Footprint Considerations

Although biodesulfurization reduces the energy required for sulfur removal, the overall environmental impact depends on the energy sources used to power the reactors. If the process is integrated with renewable energy sources like solar or wind, its carbon footprint is minimal. However, if biodesulfurization is powered by fossil fuels, the carbon savings from reduced sulfur emissions may be offset by the CO_2 generated during the energy production process. To maximize environmental benefits, biodesulfurization should be implemented alongside energy-efficient technologies or in regions where renewable energy sources are abundant.

Biodesulfurization also promotes the sustainable use of natural resources. Coal, despite being a carbon-heavy fuel, remains a critical energy source globally. By enhancing the quality of coal through sulfur removal, biodesulfurization allows coal to be used in a cleaner and more environmentally responsible manner. Furthermore, the process provides a pathway for recycling certain byproducts, such as elemental sulfur, which can be repurposed in other industries, contributing to a circular economy. Biodesulfurization presents a more environmentally sustainable option for sulfur removal from coal compared to conventional methods. Its ability to lower sulfur emissions, reduce energy consumption, and minimize hazardous byproducts makes it an attractive solution in the context of global efforts to reduce air pollution and mitigate climate change. However, successful implementation requires careful management of wastewater, biofilm formation, and the overall energy use to fully realize its environmental benefits.

7.7 Challenges and Future Perspectives

While biodesulfurization presents a promising alternative for removing sulfur from coal and reducing associated environmental impacts, several challenges must be addressed to enhance its efficiency and applicability on an industrial scale. Overcoming these challenges will be crucial for the future development and implementation of biodesulfurization technologies, ensuring their integration into sustainable energy practices.

The effectiveness of biodesulfurization largely depends on the microbial species involved and their interactions within the reactor. Identifying and cultivating the most effective microbial consortia that can thrive under varying environmental conditions is essential. Research is needed to explore the synergistic relationships among different sulfur-oxidizing and sulfate-reducing bacteria, aiming to enhance their metabolic efficiency. Moreover, the stability of these consortia over extended periods and under varying operational conditions poses a challenge. Advances in genetic engineering and synthetic biology may allow the development of tailored microbial strains that exhibit improved desulfurization capabilities, enhancing the robustness of the process.

Designing reactors that can efficiently accommodate the unique requirements of biodesulfurization is crucial for its practical application. Existing reactor designs may need to be modified to optimize conditions for microbial growth, mass transfer, and nutrient availability. Scaling up from laboratory-scale experiments to industrial-scale operations often presents significant challenges, including maintaining uniform conditions throughout larger volumes and managing byproduct accumulation. Innovative reactor designs, such as modular systems or continuous-flow reactors, could facilitate better control of process parameters and improve overall efficiency. Additionally, integrating biodesulfurization with existing coal processing facilities could lead to more efficient use of resources and infrastructure.

The economic viability of biodesulfurization compared to traditional sulfur removal methods is a significant consideration for widespread adoption. While biodesulfurization has lower energy and chemical input costs, the initial investment in reactor technology and the ongoing costs associated with microbial cultivation, maintenance, and nutrient supply can be considerable. To enhance competitiveness, further research is needed to reduce operational costs through process optimization, waste recycling, and recovery of valuable byproducts like elemental sulfur. Additionally, establishing financial incentives or regulatory frameworks that promote cleaner technologies can support the transition to biodesulfurization in coal utilization.

The use of microbial species in industrial applications raises regulatory and safety concerns, particularly regarding the potential release of genetically modified organisms (GMOs) into the environment. Regulatory frameworks must be developed to assess the safety and ecological impact of using specific microorganisms for biodesulfurization. Furthermore, ensuring that operational practices minimize the risk of contamination or unintended effects on local ecosystems is crucial for gaining public acceptance. Collaborative efforts between researchers, industry stakeholders, and regulatory bodies can help establish guidelines and best practices for the safe application of biodesulfurization technologies.

To maximize the environmental benefits of biodesulfurization, integrating the process with renewable energy sources is essential. Using renewable energy to power biodesulfurization reactors can further reduce the carbon footprint associated with coal utilization. This integration can also help create a more resilient energy system that relies on sustainable practices. Research into hybrid systems that combine biodesulfurization with other renewable technologies, such as bioenergy or waste treatment, may lead to innovative solutions that enhance overall resource efficiency and reduce reliance on fossil fuels.

The successful adoption of biodesulfurization technologies also depends on public perception and acceptance. As awareness of environmental issues and climate change grows, there is an increasing demand for sustainable and eco-friendly practices. Public education initiatives that highlight the benefits of biodesulfurization, including its potential to reduce emissions and improve air quality, can foster acceptance. Engaging with communities and stakeholders to address concerns and promote the advantages of cleaner technologies can help build support for biodesulfurization initiatives. Continued research is essential to overcome existing challenges and enhance the efficiency of biodesulfurization processes. Areas of focus include exploring the genetic and metabolic pathways of key microbial species, optimizing nutrient formulations, and investigating novel reactor designs. Additionally, research into the long-term sustainability of biodesulfurization practices, including monitoring for potential ecological impacts, will be crucial for ensuring the responsible implementation of these technologies. Interdisciplinary collaboration between microbiologists, chemical engineers, environmental scientists, and industry practitioners will drive innovation and accelerate the development of biodesulfurization as a viable solution for sustainable coal utilization.

7.8 Conclusion

Biodesulfurization presents a promising and sustainable alternative for reducing sulfur emissions from coal, addressing critical environmental challenges associated with fossil fuel combustion. By leveraging the metabolic capabilities of specialized microorganisms, this process not only enhances the quality of coal but also mitigates the harmful effects of sulfur dioxide (SO_2) emissions, which contribute to air pollution and acid rain. The design and optimization of reactors play a pivotal role in maximizing the efficiency of biodesulfurization, with various reactor types offering distinct advantages and challenges that must be carefully managed.

As the global focus shifts toward cleaner energy solutions, addressing the challenges of microbial stability, reactor scalability, and economic competitiveness will be crucial for the successful implementation of biodesulfurization technologies. Future research should prioritize the optimization of microbial consortia, the integration of renewable energy sources, and innovative reactor designs to enhance process efficiency. Additionally, fostering public awareness and acceptance will be essential to promote the adoption of biodesulfurization as a viable solution for sustainable coal utilization.

In conclusion, with continued advancements in reactor technology and a commitment to overcoming existing challenges, biodesulfurization has the potential to significantly contribute to the transition toward cleaner energy practices, ensuring a more sustainable and environmentally responsible future for coal-based energy production.

References

1. Kotelnikov V, Saryglar CA, Chysyma R (2020) Microorganisms in coal desulfurization. Appl Biochem Microbiol 56:521–525
2. Xu J, Liu X, Song C, Du Z, Wang F, Luo J, Chen X, Zhou A (2020) Biodesulfurization of high sulfur coal from Shanxi: optimization of the desulfurization parameters of three kinds of bacteria. Energy Sour Part A Recov Util Environ Effects 42(18):2297–2315
3. Liu F, Lei Y, Shi J, Zhou L, Wu Z, Dong Y, Bi W (2020) Effect of microbial nutrients supply on coal bio-desulfurization. J Hazard Mater 384:121324
4. Guchhait S, Biswas D, Bhattacharya P, Chowdhury R (2005) Bio-desulfurization of model organo-sulfur compounds and hydrotreated diesel—experiments and modeling. Chem Eng J 112(1–3):145–151
5. Nuhu AA (2013) Bio-catalytic desulfurization of fossil fuels: a mini review. Rev Environ Sci Biotechnol 12:9–23
6. Etemadi N, Sepahy AA, Mohebali G, Yazdian F, Omidi M (2018) Enhancement of bio-desulfurization capability of a newly isolated thermophilic bacterium using starch/iron nanoparticles in a controlled system. Int J Biol Macromol 120:1801–1809
7. Chen Z, Mu T, Yang M, Samak NA, Hao X, Jia Y, Yang G, Wen Q, Shen J, Peh S (2023) Biologically produced elemental sulfur clogging induced by thiols in gas biodesulfurization systems. Chem Eng J 462:142196
8. Kiragosyan K, Picard M, Timmers PH, Sorokin DY, Klok JB, Roman P, Janssen AJ (2020) Effect of methanethiol on process performance, selectivity and diversity of sulfur-oxidizing bacteria in a dual bioreactor gas biodesulfurization system. J Hazard Mater 398:123002
9. Gupta S, Plugge CM, Klok JB, Muyzer G (2022) Comparative analysis of microbial communities from different full-scale haloalkaline biodesulfurization systems. Appl Microbiol Biotechnol 106(4):1759–1776
10. Almenglo F, González-Cortés J, Ramírez M, Cantero D (2023) Recent advances in biological technologies for anoxic biogas desulfurization. Chemosphere 321:138084
11. Prasoulas G, Dimos K, Glekas P, Kalantzi S, Sarris S, Templis C, Vavitsas K, Hatzinikolaou DG, Papayannakos N, Kekos D (2021) Biodesulfurization of dibenzothiophene and its alkylated derivatives in a two-phase bubble column bioreactor by resting cells of Rhodococcus erythropolis IGTS8. PRO 9(11):2064
12. Guerrero RDBS, dos Santos CED, Soares LA, Zaiat M (2020) Comparison between two different fixed-bed reactor configurations for nitrogen removal coupled to biogas biodesulfurization. Biochem Eng J 162:107716
13. de Rink R, Gupta S, de Carolis FP, Liu D, ter Heijne A, Klok JB, Buisman CJ (2021) Effect of process conditions on the performance of a dual-reactor biodesulfurization process. J Environ Chem Eng 9(6):106450

14. Ishfaq HA, Banerjee A, Qamar S (2021) Bio-desulfurization of coal using biotechnological approach, making coal a less harmful fuel. In: Clean Coal technologies: beneficiation, utilization, transport phenomena and prospective. Springer, Cham, pp 161–179
15. Kadhum AT, Albayati TM (2022) Desulfurization techniques process and future challenges for commercial of crude oil products. In: AIP conference proceedings. AIP Publishing
16. Ahmad A, Zamzami MA, Ahmad V, Al-Thawadi S, Akhtar MS, Khan MJ (2023) Bacterial biological factories intended for the desulfurization of petroleum products in refineries. Fermentation 9(3):211
17. Sousa JOP, Neves RP, Sousa SF, Ramos MJ, Fernandes PA (2020) Reaction mechanism and determinants for efficient catalysis by DszB, a key enzyme for crude oil bio-desulfurization. ACS Catal 10(16):9545–9554
18. Zhao D, Sun P-P, Ai C-M, Mu X-Z (2023) Study on the experiment and reaction kinetics of sulfur removal from coal by microorganisms. Front Microbiol 14:1184253
19. Jatoi AS, Aziz S, Soomro SA (2021) Biological assisted organic sulfur removal from low rank indigenous coal using airlift bioreactor. Bioprocess Biosyst Eng 44:417–427
20. Li W, Zhang M, Kang D, Chen W, Yu T, Xu D, Zeng Z, Li Y, Zheng P (2020) Mechanisms of sulfur selection and sulfur secretion in a biological sulfide removal (BISURE) system. Environ Int 137:105549

Chapter 8
Kinetic and Modeling Studies of Biodesulfurization of Coal

Abstract The current chapter is the overview of the kinetics and modeling of biodesulfurization process, which covers bioprocess kinetic of sulfur removal from coal. The effects of various parameters such as temperature ranges, pH, and incubation time on coal biodesulfurization have been studied. As far as biodesulfurization of coal was concerned, it totally depends upon microbial growth. Such growth rate was measured through different models; on behalf of growth kinetics, different models were studied.

8.1 Introduction

Numerous heterocyclic organic sulfur compounds can be found in fossil fuels (coal and petroleum). It is well known that DBT and its related derivatives, which contain a significant amount of organic sulfur compounds (OSCs) in coal, are insurmountable against hydrodesulfurization (HDS). There are important developments in coal's biocatalytic desulfurization (BDS), a technique that could replace HDS. In fact, it is advised to desulfurize fuel oil by selectively removing sulfur from OSCs without damaging the hydrocarbon backbone of the substance. The calorific value of the fuel after BDS will be kept since the useful carbon of the fuel in the oil is retained and won't be lost in the water phase in contact with the oil in a water-soluble state. As a model heterocyclic organosulfur molecule in the majority of BDS investigations, DBT has received widespread acceptance Mathematical modelling has a favorable effect on comprehending the relationships between distinct variables and on offering practical ways to improve a number of factors. The fundamental dynamics of productivity changes over time are provided by the Monod model (Jatoi et al. 2016a, Jatoi et al. 2016b, Mahar et al. 2018) using numerical techniques to solve the kinetic equation. In comparison to solving differential equations to optimize process parameters, numerical approaches have several advantages (Jatoi et al. 2018, Jatoi et al. 2019, Mahar et al. 2017). A few studies that are connected to this topic, nevertheless,

A. S. Jatoi, *Biological Removal of Sulphur from Coal*, Green Energy and Technology, https://doi.org/10.1007/978-981-96-8628-5_8

concentrate on the kinetics of DBT degradation and 2-HBP formation and elucidate the process of organic sulfur elimination. The primary goal of the current study is to use the novel *Rhodococcus* sp-SL-9 to examine the effects of various parameters on the biodesulfurization of DBT. Furthermore, batch DBT degrading processes were explored using kinetic modelling and simulation. The distribution of exponents has several uses in the sciences, particularly in science biology. However, this is not the only sector where this dissemination is used; it is also used in science pathology and other areas. Regression exponent modelling can be used to describe data gathered in this way. It directly makes it easier for the researcher to watch for and analyze any changes.

8.1.1 Phases of Microbial Growth

Microbial growth occurs in distinct phases: lag, exponential (log), stationary, and death. During the **lag phase**, microorganisms adapt to their environment, synthesizing enzymes and preparing for cell division. Although cell numbers remain constant, this phase is critical for establishing metabolic processes necessary for growth. The duration of the lag phase depends on factors like inoculum size, medium composition, and environmental conditions.

In the **exponential phase**, cells divide at their maximum rate, leading to rapid population growth. Nutrient availability and optimal environmental conditions drive this phase, characterized by constant doubling time. Following the exponential phase, the **stationary phase** occurs as resources become limited and waste accumulates, resulting in a balance between cell division and death. Eventually, the **death phase** begins, marked by a decline in viable cells due to resource depletion and toxic byproducts. These phases collectively describe the life cycle of microbial populations and guide bioprocess optimization.

8.1.2 Lag Phase

In the biodegradation of dibenzothiophene (DBT) by microorganisms, the **lag phase** represents the period where microbial cells acclimate to the presence of this sulfur-containing compound as their primary substrate. During this phase, microbes synthesize specific enzymes, such as monooxygenases and desulfurases, required to initiate the breakdown of DBT. Since DBT is a heterocyclic aromatic compound with a stable structure, its biodegradation demands specialized metabolic pathways. The length of the lag phase depends on factors like the microbial strain, the concentration of DBT, and environmental conditions such as temperature and pH.

Additionally, during this phase, microbial cells may upregulate genes associated with sulfur metabolism to adapt to DBT utilization. The presence of co-metabolites or inducers, such as simpler sulfur compounds, can shorten the lag phase by priming the microorganisms for DBT degradation. Understanding and minimizing the lag phase in DBT biodegradation is essential for enhancing the efficiency of desulfurization processes, which are critical in reducing sulfur emissions from fossil fuels.

8.1.3 The Exponential Phase

The exponential phase is the second stage of growth that can be seen in a batch system. The exponential phase is defined by a time of exponential growth, which is the fastest growth achievable in the batch system's current conditions. The quantity of cells in the culture at any given time determines the rate at which the cells develop during exponential growth. This idea can be expressed theoretically and mathematically in a variety of ways. The log phase, sometimes referred to as the exponential phase of microbial growth, is distinguished by steady and quick cell division. Microorganisms develop at their fastest pace during this phase when the right conditions are present, such as enough nutrients, a suitable pH, and a favorable temperature. The log phase, sometimes referred to as the exponential phase of microbial growth, is distinguished by steady and quick cell division. The population size doubles at regular intervals, following the equation.

where N(t) is the population size at time tt, N_0 is the initial population, and μ is the specific growth rate.

This stage is essential for industrial processes, where optimizing microbial activity is frequently the aim, due to the exponential rise in cell population. The exponential phase in biodegradation or bioprocessing is the time when microorganisms are most effective in metabolizing substrates, like pollutants or feedstocks. There is a high level of enzymatic activity and complete substrate conversion by metabolic pathways. For instance, pollutants like dibenzothiophene are actively broken down into simpler chemicals by hydrocarbon-degrading bacteria throughout their exponential phase. This stage is crucial for keeping an eye on and adjusting bioreactor conditions to guarantee the intended result, like the removal of pollutants or the synthesis of new products.

8.1.4 The Stationary Phase

The stationary phase is the third stage of growth. Even in the absence of net growth, cells continue to divide and expand throughout the stationary phase. In other words, the death of an equal number of cells balances growth. There are several reasons why a batch culture could go into a stagnant phase. The depletion of a vital nutrient or the carbon and energy supply is a common explanation. The depletion of a carbon source does not necessarily mean that all growth stops. The reason for this is because dying cells can lyse and release nutrients. The growth that takes place on dead cells is known as endogenous metabolism. Endogenous metabolism occurs throughout the growth cycle, but it is most noticeable during the stationary phase, when growth is measured by the amount of oxygen consumed or the amount of carbon dioxide evolved. Because of this, many growth curves, such as the one shown in Fig. 3.6, actually show some growth during their stationary phase. This growth, once more, happens after the substrate has been used and represents the utilization of dead cells as a source of energy and carbon. Waste products accumulate to the point where they start to impede cell growth or become poisonous to cells, which is a second explanation for why a stationary phase may be seen. Usually, only cultures with a high cell density experience this. Because certain components are easier for cells to manufacture than others, growth in the stationary phase is imbalanced regardless of the source of cell entry into this phase.

The stationary phase in the biodegradation of dibenzothiophene (DBT) occurs when microbial growth slows down and approaches a dynamic equilibrium. During this phase, the rate of microbial cell division is approximately equal to the rate of cell mortality due to resource restrictions, such as diminished DBT availability or the accumulation of hazardous byproducts. In DBT biodegradation, the stationary phase is important because it shows how well the system can support microbial activity under limited circumstances. During this stage, enzymatic activity might still be present, but it would be slower, allowing DBT to partially degrade.

Moreover, in this phase, microorganisms often switch to secondary metabolic strategies to cope with the stress of limited resources and waste buildup. Some microbial strains may exhibit stress tolerance mechanisms, such as upregulating stress response genes or altering membrane composition to resist toxicity. Since interventions like nutrition supplementation, pH control, or byproduct removal can help extend microbial activity and enhance DBT degradation efficiency, it is essential to understand the dynamics of the stationary phase in order to optimize biodegradation systems.

8.1.5 The Death Phase

The death phase in the biodesulfurization of coal refers to the stage in the microbial life cycle where the microorganisms, which are responsible for sulfur removal from coal, begin to die off. This phase occurs after the initial growth and active sulfur reduction phases, typically due to nutrient depletion, accumulation of toxic byproducts, or unfavorable environmental conditions. As the microbes deplete the available sulfur compounds and other nutrients in the system, they enter a phase of reduced metabolic activity, leading to cell death.

During the death phase, the microbial population sharply declines, and the rate of biodesulfurization slows or ceases. The death of these microbes can limit the overall efficiency of the desulfurization process, especially if conditions such as pH, temperature, or oxygen levels are not optimized. In many biotechnological processes, including biodesulfurization, finding ways to extend the active phase or rejuvenate microbial populations can be key to improving efficiency and ensuring that the process remains viable over longer periods of time.

8.2 Kinetic Modeling of Cell Growth Modeling and DBT Degradation

Kinetic modeling is a crucial aspect of understanding and optimizing microbial processes involved in the biodesulfurization of coal. It provides insights into the dynamics of microbial growth, substrate utilization, and product formation, enabling the design and operation of efficient bioreactor systems. For desulfurization processes targeting dibenzothiophene (DBT) degradation, kinetic models can help elucidate the relationship between microbial growth and the rate of sulfur compound removal, facilitating process optimization and scalability. Kinetic models play a critical role in scaling up biodesulfurization processes. They allow researchers to:

- Predict microbial growth and DBT degradation rates under different operational conditions, such as pH, temperature, and agitation speed.
- Optimize bioreactor design and operating parameters for maximum efficiency.
- Estimate the time required to achieve target sulfur removal levels.

  paving the way for more efficient and sustainable biodesulfurization technologies.

8.2.1 Growth Kinetic of Rhodococcus *spp.*

Rhodococcus spp. is a genus of Gram-positive, aerobic, and non-motile bacteria widely known for its ability to degrade hydrocarbons and remove sulfur compounds from fossil fuels. The desulfurization process, mainly facilitated by specific metabolic pathways, plays a significant role in bioremediation and the petroleum industry. Understanding the growth kinetics of *Rhodococcus* spp. in sulfur removal provides insights into optimizing bioprocesses for effective sulfur reduction. The growth kinetics of *Rhodococcus* spp. in sulfur removal is influenced by substrate concentration, environmental factors, and nutrient availability. Growth can be described by the following kinetic parameters: During the lag phase, bacteria acclimatize to the sulfur-containing medium. The duration depends on factors such as pre-culturing conditions, sulfur concentration, and microbial adaptation. *Rhodococcus* spp. grows rapidly in the presence of DBT, utilizing it as a sulfur source while sustaining hydrocarbon metabolism. The growth follows the Monod equation.

Growth slows as sulfur compounds deplete, metabolic byproducts accumulate, or essential nutrients become limited. In this phase, the bacterial population maintains itself, but sulfur removal efficiency declines. If sulfur sources are exhausted or toxic byproducts accumulate, *Rhodococcus* spp. enters the decline phase. Cells experience stress, reducing enzymatic activity and overall desulfurization capacity.

Biological growth of *Rhodococcus* spp. (SL-9), *with the concentration,*

$$\frac{dC_x}{dt} = \mu C_x \tag{8.1}$$

where $\frac{dC_x}{dt}$ is known as growth rate, C_x is cell concentration (g/L) at time of incubation t (h), and μ represents the specific growth rate of *Rhodococcus* spp. (SL-9) in (h^{-1}).By applying Eq. (8.1) and incorporating the term $\left[1-\frac{C_x}{C_x^{\max}}\right]$, the resulting growth rate $\frac{dC_x}{dt}$ can be expressed as follows:

$$\frac{dC_x}{dt} = \mu C_x\left[1-\frac{C_x}{C_x^{\max}}\right] \tag{8.2}$$

Equation (8.2) represents a differential form related to the logistic model equation, where C_x^{max} denotes the maximum biomass concentration during the exponential phase. The term $\left[1-\frac{C_x}{C_x^{\max}}\right]$ limits the growth rate as the population increases.

Where Cx is said to be smaller than that of C_x^{max}, i.e., $Cx <<< C_x^{max}$, the term value is about 1, which does not impact the rate of growth, but as far as increase in Cx takes the value of C_x^{max}, i.e., C_x C_x^{max}, the term value tends to zero, which makes growth rate, $\frac{dC_x}{dt}$, almost equal to zero; that is said to be the case in stationary phase.

The integration of Eq. (8.2) is performed using the initial condition $t = 0$, $C_x = C_{xo}$, resulting in the logistic growth model, represented by Eq. (8.3). This model is widely used to describe the growth of biological populations [1, 2].

$$C_x(t) = C_x = \frac{C_{x_o}e^{\mu t}}{\left[1-\left(\frac{C_{x_o}}{C_x^{\max}}\right)\left(1-e^{\mu t}\right)\right]} \tag{8.3}$$

Equation (8.3) was derived using non-linear regression, applying the Levenberg-Marquardt algorithm for parameter estimation [3] as shown in Fig. 8. The estimated parameters correspond to different initial DBT concentrations. Notably, the specific development rate decreased as the initial DBT concentration increased (Fig. 8), indicating both the toxic and inhibitory effects of DBT. However, the growth curve in Fig. 8 clearly shows that no lag phase occurs at any initial DBT concentration. Despite its toxicity, this highlights the adaptability of the biological desulfurizer SL-9 to the refractory DBT under investigation. This adaptability may be linked to the previous extraction of SL-9 (2.64%) from high-sulfur coke samples. Furthermore, the analysis revealed that the stationary phase duration increased slightly with a rise in the initial DBT concentration (Fig. 8).

8.2.2 Kinetic Model for BDS of DBT

In the case of single-substrate-based limited process, that model presented by Monod (1949) is mostly utilized for explaining the rate of consumption of substrate (Eq. 8.4):

$$q = \left(q_{\max} S\right) / \left(K_s + S\right) \tag{8.4}$$

On the other hand, *S* stands for substrate concentration, and *q* for the rate of particular substrate consumption. In order to explain the effect of substrate concentration (*S*) on substrate removal (-dS/dt) by microbe concentration (Cx), Lawrence and McCarty (1970) adapted the Monod equation:

$$ds / dt = -q\, C_x \tag{8.5}$$

By using yield coefficient $Yx_{/s}$,

$$Y_{x/S} = \left(C_x - C_{x\circ}\right) / \left(S\circ - S\right) \tag{8.6}$$

Rearranging Monod equation in terms of microbial growth by adding the term *Yx/s*,

$$Y_{x/s} = \left(\frac{dC_X}{dt}\right) / \left(\frac{dS}{dt}\right) \tag{8.7}$$

By rearrangement of Eq. (8.7),

$$\frac{dS}{dt} = -\left(1 / Y_{\frac{x}{s}}\right)\left(\frac{dC_x}{dt}\right) \tag{8.8}$$

whereas the BDS specific rate $q(h^{-1})$ can be associated with the rate of specific growth μ (h^{-1}) [4]:

$$q = \frac{\mu}{Y_{x/s}} \tag{8.9}$$

The rate of substrate consumption is a key indicator of microbial performance. In this study, a model based on a specific equation was used to simulate changes in substrate concentration over time. Figures 8.1 and 8.2 present the differential form

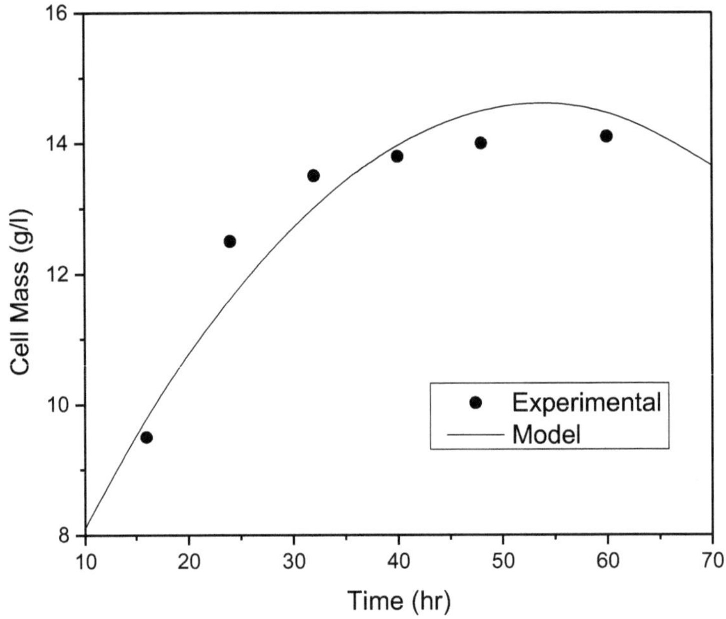

Fig. 8.1 Comparison of experimental and model results of cell mass growth at different DBT concentration

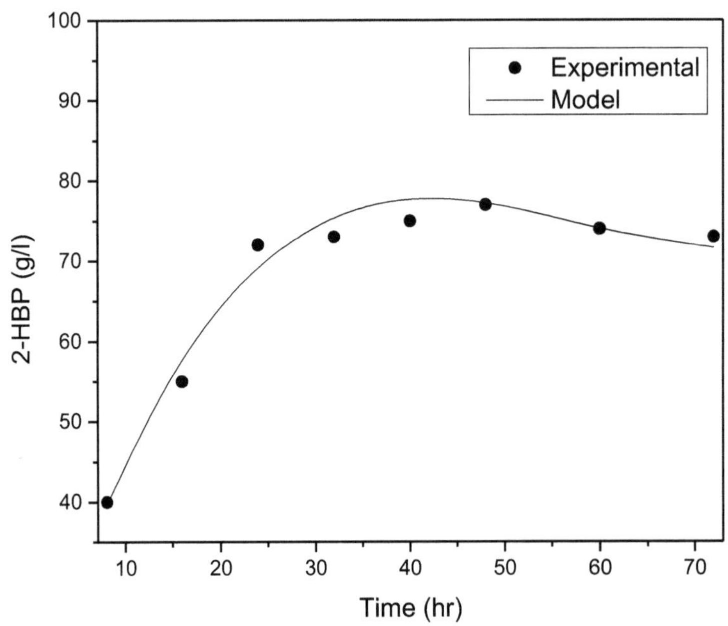

Fig. 8.2 Experimental and model results of 2-HBP production

of the biodynamic logic model (Eq. 8.2). The BDS batch processes were simulated under growing cell conditions using various initial concentrations of DBT. Fig. 8 demonstrates how the model predicts the DBT concentration curve and estimates DBT removal based on different initial concentrations. This highlights the simplicity and effectiveness of the Monod model.

The highest specific desulfurization rate (q) recorded during the BDS batch process, with an initial DBT concentration of 3.5 mM, was 0.784 h^{-1}. It was observed that the specific desulfurization rate (q) increases as the initial DBT concentration rises. Additionally, Fig. 10 clearly shows that, irrespective of the initial DBT concentration, the maximum desulfurization conversion rate occurs at approximately 70 h of incubation. After this peak, the rate gradually increases before declining.

The approximate records of DBT (dibenzothiophene) elimination for batch processing with initial concentrations of 2, 2.5, 3, and 3.5 mM revealed eliminations of 92.3%, 76.4%, 65.6%, and 73.6%, respectively, after 70 h. These values further increased to 96.4%, 94.2%, 82.5%, and 82.2%, respectively. For batches with initial concentrations of 2 and 2.5 mM, near-complete elimination of DBT was observed. However, for initial concentrations of 3 and 3.5 mM, approximately 92% and 86.5% elimination were achieved after 8 days. The reduction in desulfurization rates over time is attributed to feedback inhibition caused by sulfated and toxic byproducts such as 2-hydroxybiphenyl (2-HBP).

Interestingly, the peak desulfurization activity was observed at 70 h of growth, indicating maximum enzyme activity. This suggests that cells harvested at this time can serve as highly efficient biocatalysts for desulfurization when using resting cells. Resting cells are preferred over growing cells for biodesulfurization (BDS) processes because they yield higher desulfurization efficiency. The age of the biocatalyst plays a crucial role in determining BDS efficiency, as studies show that optimal activity is linked to the cell age.

Research by Ohshiro et al. (1996) and Konishi et al. (1997) supports the use of resting cells to overcome cofactor regeneration challenges, such as NADH, associated with the first three oxidation steps of the 4S pathway. The efficiency of these enzymatic steps is highly dependent on cell age, as demonstrated in studies on *Pseudomonas putida* CECT5279. Enzyme DszB, produced consistently during the growth phase, shows peak activity early in the exponential phase but rapidly declines thereafter, limiting efficient conversion of HBPS to HBP.

Further studies highlight that the highest DBT elimination and 2-HBP production rates occur during the logarithmic growth phase, regardless of the initial DBT concentration. The BDS yield improves significantly with higher cell concentrations, although excessive DBT concentrations can inhibit cell growth and desulfurization activity.

This study's findings show that while the specific growth rate (μ, ranging from 0.06435 to 0.04741 h^{-1}) is relatively unaffected by increasing initial DBT concentrations, the specific desulfurization rate (q, ranging from 0.1408 to 0.8784 h^{-1}) increases with higher concentrations (100–1000 ppm). The Monod model parameters for BDS, including the maximum desulfurization rate ($q_{max} = 0.7818\ h^{-1}$) and saturation constant ($K_s = 588$ mg/L), further validate these findings.

The collected data demonstrate that *Rhodococcus* sp. SL-9 is well adapted to high concentrations of heterologous organisms and refractory sulfur compounds, making it a promising candidate for the biodesulfurization of real oil supplies with high initial sulfur concentrations.

8.3 Modeling of Population Growth as a Function of Incubation Time

Modeling population growth as a function of incubation time is fundamental in understanding microbial dynamics. It involves using mathematical models to predict microbial population changes over time under specific conditions. These models account for the lag phase, exponential phase, stationary phase, and death phase, which reflect different microbial responses to environmental conditions. Accurate modeling aids in optimizing incubation times for industrial processes, including fermentation and biodegradation.

The relationship between incubation time and microbial growth provides insights into growth rate, population density, and system limitations. Factors such as nutrient availability, temperature, and pH influence these dynamics, making models essential for simulating real-world scenarios. Researchers and industries utilize these models to design efficient bioprocesses, ensuring maximum productivity and cost-effectiveness.

8.4 Microbial Growth Kinetics

Microbial growth kinetics describes the rates at which microbial populations grow under varying conditions. Key parameters include the specific growth rate, yield coefficient, and doubling time, all of which are influenced by environmental and nutritional factors. Understanding these kinetics is crucial for scaling up biological processes such as wastewater treatment, biofuel production, and pharmaceuticals.

Growth kinetics also play a significant role in managing microbial competition in mixed cultures. By analyzing growth curves and kinetic parameters, engineers can optimize conditions to favor desirable microbes, enhancing process efficiency and sustainability. These insights support the development of predictive models for various bioprocess applications.

8.5 Exponential Growth Model

The exponential growth model is a simple representation of microbial population growth during the exponential phase, where resources are abundant. In this phase, the population doubles at regular intervals, leading to rapid increases in cell numbers. The model is described mathematically by $N(t) = N_0e^{\mu t}$, where N_0 is the initial population, μ is the specific growth rate, and tt is time.

While the exponential growth model provides a clear understanding of rapid population increases, it has limitations as it assumes unlimited resources and ignores environmental constraints. Despite these limitations, it serves as a fundamental tool for studying microbial behavior and designing initial stages of industrial processes.

8.6 Logistic Growth Model

The logistic growth model improves upon the exponential model by incorporating resource limitations and carrying capacity. This model describes microbial growth as initially rapid, followed by a deceleration as resources deplete, eventually reaching a steady-state population size. The equation $N(t) = \frac{K}{1 + \frac{(K - N_0)}{N_0} e^{-\mu t}}$ integrates carrying capacity KK and considers environmental constraints.

This model is particularly valuable in predicting long-term microbial dynamics in closed systems, such as bioreactors. It offers insights into the effects of competition and resource depletion, guiding optimization in biotechnological and environmental applications.

8.7 Some of the Classical Kinetic Models Applied in BDS Studies

Classical kinetic models such as Monod, Contois, and Tessier models are widely used in biodegradation and detoxification studies (BDS). The Monod model, for example, relates microbial growth rate to substrate concentration, effectively predicting the dynamics of biodegradation. These models help optimize bioprocesses by understanding substrate utilization patterns and growth efficiencies.

These models also provide frameworks for integrating microbial growth with environmental factors, improving predictions for complex systems. Their application in BDS aids in achieving efficient waste treatment, pollutant degradation, and resource recovery, making them indispensable in environmental engineering.

8.8 Factors Affecting the Rate of Microbial Growth

8.8.1 *Effect of Temperature*

Temperature significantly impacts microbial growth by influencing enzyme activity and membrane fluidity. Optimal temperatures maximize growth rates, while extreme temperatures can inhibit or kill microbial cells. Mesophilic microbes thrive at moderate temperatures, whereas thermophiles and psychrophiles are adapted to extreme conditions.

In industrial and environmental applications, maintaining the right temperature is crucial for desired microbial activity. For instance, thermophilic conditions are exploited in composting and anaerobic digestion to enhance decomposition rates and pathogen reduction.

8.8.2 *Effect of pH*

Microbial growth is sensitive to pH, as it affects enzyme function and cellular processes. Most microbes prefer a neutral pH, though acidophiles and alkaliphiles thrive in extreme pH environments. Drastic pH changes can disrupt cell walls, denature proteins, and inhibit growth.

In bioprocessing, controlling pH is critical to ensure optimal conditions for microbial activity. Wastewater treatment and fermentation industries often use buffer systems to stabilize pH, enhancing process efficiency and productivity.

8.8.3 *Effect of Oxygen*

Oxygen availability dictates microbial metabolism, dividing microbes into aerobes, anaerobes, and facultative anaerobes. Aerobic microbes require oxygen for energy production, while anaerobes rely on alternative electron acceptors. Facultative anaerobes can switch between both modes depending on oxygen levels.

Understanding oxygen's role helps optimize bioreactor designs and environmental conditions for microbial processes. For instance, oxygen supply in aerobic wastewater treatment must be controlled to maintain effective decomposition rates without over-aeration.

8.9 Conclusion

Microbial growth and kinetics are central to understanding and optimizing biological processes. From modeling population growth to analyzing the effects of environmental factors, these insights drive advancements in biotechnology, environmental management, and industrial applications.

Continued research and application of microbial growth models enable the development of sustainable solutions for global challenges. By leveraging these models, industries can enhance productivity, reduce costs, and promote environmental stewardship.

References

1. Martin AB, Alcon A, Santos VE, Garcia-Ochoa F (2004) Production of a biocatalyst of Pseudomonas putida CECT5279 for Dibenzothiophene (DBT) biodesulfurization for different media compositions. Energy Fuel 18:851–857
2. Martin AB, Alcon A, Santos VE, Garcia-Ochoa F (2005) Production of a biocatalyst of Pseudomonas putida CECT5279 for DBT biodesulfurization: Influence of the operational conditions. Energy Fuel 19:775–782
3. Marquardt DW (1963) An algorithm for least-squares estimation of nonlinear parameters. J Soc Ind Appl Math 11:431–441
4. Nassar HN, Deriase SF, El-Gendy NS (2017) Kinetic modeling and simulation of a batch biodesulfurization process by Rhodococcus erythropolis HN2. Pet Sci Technol 35:1995–2002

Chapter 9
Challenges and Perspectives of Biodesulfurization of Coal

Abstract Biodesulfurization of coal, utilizing microbial techniques to eliminate sulfur compounds, is a sustainable and environmentally benign alternative to traditional desulfurization technologies. This method shows promise in alleviating environmental issues including acid rain and air pollution; however it encounters considerable obstacles. This encompasses restricted efficacy in eliminating organically bound sulfur, sluggish microbial activity, and heterogeneity in coal content that influences treatment results. Progress in biotechnology, including genetic engineering and omics technologies, as well as novel bioreactor designs and pretreatment techniques, is crucial for tackling these difficulties. Ongoing research and development indicate that biodesulfurization may serve as an effective alternative for cleaner coal consumption and sustainable energy production.

Commercialization of biodesulfurization of coal needs more attention with respect to current available methods for sulfur remediation techniques. As far as clean coal technologies are available for coal one of the major issue. In this regard the current chapter provides an opportunity to address the current and future perspective in this chapter regarding biodesulfurization of coal.

9.1 Introduction

The utilization of coal as a principal energy source presents considerable environmental issues, notably the release of sulfur oxides (SO_2) during combustion [1]. These emissions result in acid rain, respiratory health complications, and worldwide air pollution [2]. Conventional techniques for sulfur elimination, including chemical and physical desulfurization, are energy-demanding, expensive, and frequently ineffective in removing organically bonded sulfur [3]. In this context, biodesulfurization, which utilizes microorganisms to selectively extract sulfur from coal, has emerged as a potential, environmentally benign, and economical option. Nonetheless, its adoption and extensive implementation encounter numerous technological and practical obstacles [4].

A. S. Jatoi, *Biological Removal of Sulphur from Coal*, Green Energy and Technology, https://doi.org/10.1007/978-981-96-8628-5_9

A primary problem in biodesulfurization is the efficacy of microbial processes [5]. The microorganisms employed, including *Acidithiobacillus ferrooxidans* and *Rhodococcus* species, are proficient in eliminating pyritic sulfur but exhibit reduced efficacy in addressing organically bonded sulfur, which is more intricate and intricately integrated within the coal matrix. Furthermore, the process is frequently protracted, necessitating prolonged treatment durations, which may impede its industrial viability [6]. Enhancing microbial activity, creating genetically modified strains with superior functions, and refining bioreactor designs are crucial for addressing these constraints.

A further considerable problem resides in the intrinsic characteristics and variability of coal. Variations in coal composition, including sulfur content, mineral impurities, and moisture levels, influence the biodesulfurization process. Elevated ash content or compact coal structures may restrict microbial access to sulfur compounds. Moreover, transitioning from laboratory studies to industrial operations presents challenges in sustaining uniform microbial activity and process conditions across different coal varieties. These issues require extensive research to customize biodesulfurization methods to particular coal properties.

Notwithstanding these obstacles, the prospects for biodesulfurization remain optimistic [7, 8]. Innovations in biotechnology, encompassing metabolic engineering and omics technologies, are facilitating the development of more resilient and efficient microbial strains. In conjunction with advanced reactor systems and pretreatment techniques to improve coal's bioavailability, these advancements could render biodesulfurization a feasible industrial alternative. Moreover, the integration of biodesulfurization with other clean coal technologies might substantially enhance sustainable energy generation while reducing the environmental consequences of coal use.

9.2 Biodesulfurization Under Thermophilic Conditions

Thermophilic conditions are desirable for biodesulfurization because they increase the BDS rate and bring the operating temperature closer to FCC or HDS output streams. A higher temperature boosts enzymatic rates, reduces bacterial contamination, reduces oil viscosity, and facilitates molecular displacement [9]. However, because distillate fractions are frequently processed at high temperatures, using a moderate thermophile might result in some cost savings because it would be unnecessary to cool the fractions to room temperature. As a result, BDS might be more readily integrated into a refining stream after HDS [10]. Additionally, the larger mass transfer rate at high temperatures would likewise increase desulfurization activity. Obtaining microorganisms with significantly better BT and DBT desulfurization capabilities at high temperatures is beneficial for practical BDS [11]. Thermophilic forms of the 4S-pathway have also been identified and characterized, and operons encoding the enzymes in these 4S routes, such as tdsABC (Ishi Enzymes with tdsABC coding) are functional up to roughly 60 °C. It has been demonstrated that the extreme thermopile *Pyrococcus furiosus* may convert organic polysulfide to H2S by employing the reduction of sulfur as an energy source. *Paenibacillus* sp. A11–2, a thermophilic bacterium with an aerobic desulfurization characteristic that desulfurizes DBT at 60 °C, was described as the first in 1997 [12]. The strain had a distinct desulfurization gene cluster (tdsABC), whose genes shared 73%, 61%, and 52% of their sequences with the dszABC genes. It can desulfurize DBT selectively without harming the hydrocarbon matrix. This strain uses the same *Rhodococcus erythropolis* IGTS8 metabolic pathway [13]. Similar to *Rhodococcus* enzymes, *Paenibacillus* enzymes are active at temperatures between 50 and 60 °C. Thus, it is suggested to create a BDS process for crude oil at high temperatures, where mass transfer constraints are lessened and crude oil viscosity is lower. However, compared to the mesophilic cultures previously reported at the time, A11-specific 2's desulfurization activity, 0.008 mM 2-HBP/h/g DCW, was lower The complete oxidation of 149 ppm DBT (26 ppm sulfur) was catalyzed by resting cells from a thermophile strain of *M. phlei* WU-F1 in a study funded by the Japan Cooperation Center, Petroleum (JCCP). These cells also demonstrated desulfurizing ability toward naphthothiophene found in gas oil and hydrodesulfurized light gas oil at 50 and 45 °C, respectively. Additionally, this bacterium reduced the sulfur content of hydrodesulfurized gas oil by 60–70%. Similarly, *M. phlei* WU-0103 was successful in reducing 52% of the sulfur in a crude straight-run light gas oil fraction that had not been HDS-treated. Another thermophilic bacterium, *B. subtilis* WUS2B, is able to breakdown DBT entirely at 12 h and its derivatives to around 50% at 24 h by selectively cleaving C-S bonds, producing a buildup of 2-HBP. To remove the sulfur atom without causing a break in the C-C bond and hence without losing calorific value, this type of reaction would be preferable. It was discovered and given the designation bdsABC for the DBT desulfurization gene cluster in WU-S2B. Sequencing of the bds genes' DNA and amino acids in comparison to those of IGTS8 revealed a 61% homology. By using the facultative thermophilic *Mycobacterium goodii* X7B, Li et al. (2007) reported the BDS of model oil (DBT in tetradecane), where the total sulfur level of DBT in tetradecane was decreased by 99% from 200 to 2 ppm in just 24 h at 40 °C. Gun et al. (2015) observed that *Sulfolobus solfataricus* P2 was capable of desulfurizing DBT and its derivatives at 78 °C but not BT, with a DBT-BDS activity of roughly 1.23 mol 2-HBP/g DCW/h. In a biphasic system (DBT dissolved in xylene, 40% O/W), it also demonstrated good DBT-BDS capability, registering 0.34 mol DBT/g DCW/h.

9.3 Anaerobic Biodesulfurization

An appealing possibility is reductive or anaerobic biodesulfurization. Sulfate-reducing bacteria (SRB) are used in one method that has attracted a lot of interest to reductively desulfurize organic sulfur compounds [14]. These organisms possess the unusual capability of converting sulfate to H2S in order to produce energy. Hydrogen, which is more expensive than oxygen, is what propels this choice.

Desulfovibrio desulfuricans M6 [15], the extreme thermopile *Pyrococcus furiosus* [16], *Nocardioform actinomycete* FE9 [17], *Desulfomicrobium scambium*, and *Desulfovibrio longreachii* have all been used in anaerobic BDS of crude oil and its. Because it avoids the expenditures involved with aeration, has the benefit of releasing sulfur as a gas and does not release sulfate as a byproduct that needs to be disposed of by suitable treatment; the desulfurization of petroleum under anaerobic conditions would be appealing. Due to the untacking of C-C bonds and the similarity of the reaction pattern to HDS, it retains its caloric value. There is little oxidation of hydrocarbons into undesirable chemicals, such as pigmented, acidic, and/or gum-forming products, when anaerobic conditions are present. Additionally, the aerobic technique produces diluted sulfate as a byproduct that also needs to be eliminated, whereas the anaerobic route produces H2S, which may be treated by already-existing refinery desulfurization plants (e.g., Claus process). Furthermore, growth under anaerobic conditions proceeds slowly, especially when organic molecules (like thiophenes) are involved in the conversion [18]. Growth under anaerobic conditions does not produce the toxic metabolites, such as 2-HBP, which is also involved in the formation of viscous oil sludge ("gum") in the fuel. Thiophenes serve as the only electron acceptor in anaerobic environments, and their conversion is accompanied by microbial expansion. However, the aerobic approach has several significant disadvantages from the perspective of the process [19].

Aerobic bacteria employ sulfur in their assimilatory metabolism. Given that biomass has a sulfur content of just about 0.03 weight percent, the aerobic method has a high biomass yield per mole of sulfur eliminated. In contrast, accumulation of wasted biomass won't happen in a microbial process where dissimilatory sulfur metabolism is involved (Lizama et al., 1995). Anaerobic microbes require only around 10% of their energy for assimilation, whereas aerobic microorganisms need about 50% of their energy for growth. High biomass concentrations, however, hamper downstream processing because the biomass' proteins emulsify the oil/water mixture. Additionally, in emulsions with a high biomass percentage, mixing effectiveness and oxygen availability is less than ideal. The primary consideration in the early bioreactor design suitable for anaerobically converting non-polar organic sulfur compounds is the requirement of an intimate biomass-substrate interaction. To convert thiophenes, however, suitable carbon and energy sources are also required. Due to its low cost of application at a reasonably large scale and accessibility at refineries, H2 gas is the ideal option from a process standpoint as an electron donor for converting sulfur compounds (Van Houten et al., 1996). (Gary and Handwerk, 1994). Acetate and bicarbonate, which are both needed in trace levels for growth, should be added to the aqueous phase as a source of carbon. H2 and H2S are also implicated as reaction products. The anaerobic process can therefore be thought of as a three-phase system, composed of gas, water, and hydrocarbon. It is suggested to spread the hydrocarbon phase as extremely small droplets in the aqueous phase in order to maximize the availability of organic sulfur compounds. The hydrocarbon phase is saturated with H2 gas before being introduced into the bioreactor in order to be able to deliver H2 gas to the dispersion without excessive foaming. The H2 gas is then transported by the hydrocarbon phase. The availability of both non-polar organic sulfur compounds and H2 gas is maximized by combining the needs of a high specific surface area. The best equipment to create fine hydrocarbon droplets is a bioreactor system with a nozzle to distribute the hydrocarbon phase since only the hydrocarbon phase receives mixing energy. The sulfide content must be as low as feasible to prevent inhibition and to favor the conversion of organic sulfur compounds. When H2 saturation occurs, the H2S that was created dissolves in the hydrocarbon phase and is taken off. Any sulfide inhibition is prevented in this manner. The benefits of anaerobic processes stated above should serve as motivation for further study into reductive biodesulfurization. However, maintaining an anaerobic process is very challenging, and most isolated strains' particular activity for DBTs and oil feed—such as vacuum gas oil, deasphalted oil, or bitumen—has been observed to be low. The lack of identification of specific enzymes and genes responsible for anaerobic desulfurization, low reaction rates, safety and cost concerns, and other factors have prevented the discovery of anaerobic microorganisms that are efficient enough to desulfurize petroleum in a practical manner. This has prevented the development of an anaerobic BDS process.

As a result, the majority of BDS research has focused on aerobic BDS (Le Borgne and Quintero, 2003). Similar to reductive desulfurization, dissimilatory and assimilatory systems can be used to categorize oxidative or aerobic biodesulfurization. These systems produce water-soluble sulfur products in the form of oxidized compounds (sulfate ions). When reduced sulfur compounds are utilized as electron donors in energy-producing processes, dissimilatory oxidation happens. To obtain sulfur in a state suitable for the biosynthesis of organic sulfur compounds, assimilatory sulfur oxidation is used [20].

9.4 Biocatalytic Oxidation

Organosulfur compounds can be bio-oxidized either independently using batch, semi-continuous, or continuous procedures or in combination with one or more further refining processes. Either an open or a closed vessel can be used to conduct the reaction. When compared to other hemoenzymes, cytochrome C is a biocatalyst that can oxidize thiophenes and organo-sulfides and has a number of benefits. It is inexpensive, exhibits activity at high concentrations of organic solvents, exhibits activity in a pH range of 2–11, contains the hemoprosthetic group covalently linked, and is active in a pH range of 2–11. The chloroperoxidase (CPO)-catalyzed oxidation of chemically varied OSCs demonstrates the enzyme's potential for use in the management of complicated streams [21, 22]. The biggest and most persistent problem with CPO has been the inability to generate steady heterologous production of this enzyme in a suitable host. As a result, non-genetic methods have been used to address the challenges of peroxide stability, substrate partition improvement, and overproduction in large-scale applications [23]. The CPO enzyme demonstrated a turnover number of 500,000 (i.e., the maximum number of substrate molecules that can be transformed by one enzyme molecule before inactivation). Consequently, 1 g of enzyme could lower the sulfur level of 0.81 tonnes of gasoline from 500 to 30 ppm. The desulfurization of one barrel of fuel would be too expensive to be used on a big scale, though, given the current high price of enzymes. According to estimates, for an enzyme to be economically viable, its stability must be enhanced by two orders of magnitude. However, it should be remembered that a biotechnological approach would be more affordable in terms of capital expenditure and energy use [24].

9.5 Perspectives for Enhancing the Rate of BDS

The resulting BDS rate is still two orders of magnitude behind what is needed to build a commercial BDS for crude oils and refinery streams despite all the unique and effective techniques. In an effort to be compatible with a commercial application, genetic engineering was utilized to enhance the IGTS8 strain, which resulted in a 200-fold increase in DBT desulfurizing activity and shorter residence times [25]. In order to do this, changes to the IGTS8 strain's genetic makeup and growing conditions were made, and DBT desulfurization rates of about 20 mol/min/g DCW were reached. Researchers have used a variety of tactics to enhance the BDS process for the removal of sulfur from model compounds like DBT, including strain modification through gene-editing methods, in silico studies for sulfur metabolism, and in-depth knowledge of the bioprocess development phases. It has been discovered that sequential BDS of real oil feeds can increase desulfurization value. *Mycobacterium phlei* WU-0103 was used in ten cultivation rounds, each lasting 3 days, to reduce the total sulfur in light gas oil by 52% [26]. used *Gordonia* sp. IITR100 to conduct two subsequent BDS rounds, each lasting 7 days, and reported around 61% BDS of heavy crude oil with an initial S-content of 1.083 wt.% [27].

9.6 Application of Genetics in BDS

9.6.1 Implementation of Resting Cells

A number of steps make up the BDS process via 4S, including (i) growing the chosen strain in an appropriate environment to produce cells with the maximum desulfurization activity feasible and (ii) removing active cells by cells at rest (stationary). The growth medium's composition is crucial for the development of resting cells with a high density and the highest level of desulfurization activity [28]. Additionally, although a microorganism's life in the BDS process is typically brief (1–2 days), it has been increased to 8–16 days. Although the biocatalyst lifetime is 200–400 h, new designs would enable the creation and regeneration of the biocatalyst with the BDS method. Sulfate and sulfur-containing amino acids suppress the expression of the desulfurization activity of developing cells, while 2-HBP inhibits both cell development and desulfurization activity [29]. Once desulfurization was triggered, the sulfate did not significantly reduce *Rhodococcus erythropolis* D-1 activity, even at a concentration of 1 mM. Even with 2-HBP at 10 mM in the model oil system, the *R. erythropolis* KA2–5-1 resting cells' 80% activity persisted (Morio et al., 2001). *Pseudomonas delafieldii* R-8 resting cells are not significantly impacted by pH changes, although the growth and BDS of growing cells are hindered at a pH below 5. The components of the aqueous phase and the composition of the media have a significant impact on growing cells, and they require cofactors or reducing agents. However, *Pseudomonas delafieldii* R-8 resting cells were influenced by the aqueous phase and didn't require cofactors or reducing agents to attain BD. In light of all these considerations, growing cells may be preferable than resting cells in actual fuel desulfurization. Additionally, one of the issues with two-phase BDS systems is the development of stable emulsion, which makes it challenging to separate when the reaction is finished. The use of emulsion-destabilizing chemical agents, a cell-immobilized BDS process, and avoiding the creation of a stable water-in-oil emulsion in order to simplify oil recovery are a few possible solutions for this issue that has been documented. Additionally, a supplementary step will be required to solve the separation issues, adding to the process' expense. However, utilizing resting cells of *G. alkanivorans* RIPI90A that were harvested at the time of the change from the late exponential growth phase to the stationary phase [30] observed the presence of a considerable drop in emulsion-stabilizing activity. Accordingly, Mohebali et al. (2007b) proposed that altering the reactor conditions would be necessary to provide stationary-phase conditions for the breakage of the particle-stabilized emulsions. These conditions would facilitate the preparation of a stable emulsion, which would have more efficient mass transfer than in the next step, to facilitate oil and cell recovery. Use of a resting cell reaction system is common (Table 12.6). In this method, cells are pre-grown with DBT serving as the only sulfur source and an easily accessible carbon source (such as glucose). A batch reaction is carried out after cell harvesting and concentration.

9.6.2 Microbial Consortium and BDS

Despite the advancements made over the past 20 years, BDS has not yet been used on a commercial basis. This is attributable to a number of concerns with the microbial biocatalyst's stability and catalytic effectiveness, in addition to other technical issues. Most investigations on microbial desulfurization have used axenic cultures of certain bacteria. In order to gain from cooperative or synergistic microbe-microbe interactions, it is worthwhile to investigate the BDS capacities of microbial consortia [31]. Intentional mixes of Enterobacter cloacae, which lacks desulfurization genes, and the desulfurization-competent *R. erythropolis* were created in a range of ratios spanning six orders of magnitude [32]. When the specific activity for BDS was computed, taking into account the relative abundance of the desulfurization competent cells, it was revealed that mixed cultures attained values that were 200 to 1000 times greater than pure cultures as compared to growth utilizing [33] DBT as the only source of sulfur (Kayser et al., 2003). If complete populations of biocatalysts could reach this degree of enhanced activity, it could be possible to develop an affordable BDS method for petroleum. Using resting cells of various ages can increase the efficacy of a single-species culture. Additionally, it has been suggested that combining various species can improve BDS effectiveness. For instance, a diesel fuel's total sulfur level was reduced by 86% when *R. erythropolis* DS-3 and *Gordonia* sp. C-6 were combined [31]. have described a novel method for improving desulfurization and revalorization of oil sulfur compounds using tailored synthetic bacterial consortia. This innovative method was created to get around the 4S pathway intermediates' ability to block the Dsz enzymes and to facilitate the efficient creation of value-added intermediates like 2-(2-hydroxyphenyl) benzene sulfinate (HBPSi), which is challenging to produce in monocultures. According to Ismail et al. (2016), the existence of synergistic or cooperative effects between all the species present gives microbial consortia a stronger potential for a practical BDS use. In addition to desulfurizing DBT, this consortium could also desulfurize 4-methyl DBT and 4,6-dimethyl DBT. The substrate employed affected the species richness and diversity, and the 4-methyl DBT culture showed the highest levels of both. Therefore, microbial consortium applications may be more adaptable [34, 35].

9.6.3 Surfactants and BDS

In terms of its occupancy of the recoverable fossil fuel resources, which is 66.8%, coal remains one of the irreplaceable energy supplies. More than 100 power plants, more than 500 industrial furnaces, and thousands of different kilns have used coal in China up to this point. However, a significant amount of poisonous or dangerous gas, particularly SO_2, was emitted into the air as a result of the combustion of high sulfur coal, which resulted in considerable pollution and significant environmental issues. However, a lot of coal isn't being used effectively, and the coal chemical industry's growth is anticipated to make things worse. It is promoting research into fresh uses for coal pitch as a result. Reducing the sulfur content of coal before combustion is the greatest way to reduce the amount of sulfur oxides released into the atmosphere.

As a result, research in this field has received more attention in recent years. Biodesulfurization technology has proven to be one of the most promising methods for coal desulfurization. To desulfurize coal and bioleach coal pyrite, He et al. utilized the acidophilic and thermophilic strain *Acidithiobacillus caldus*. Last but not the least, the desulfurization of coal pyritic with *A. caldus* was about 47%, and the desulfurization overall was 19%. Cardona et al. desulfurized two coals from the southwest of Colombia using a collection of local microorganisms (6). According to the findings, 31–51% of total sulfur and 85–95% of pyritic sulfur were eliminated over the course of 30 days. Microbials have been used by the Central Research

Institute of Electric Power Industry in Japan since the 1990s to increase the effectiveness of desulfurization because they may adsorb on the surface of minerals and alter their physicochemical properties (7). The ability of a bacterium species (*Pseudomonas aeruginosa*) to make a biosurfactant was discovered by Fazaelipoor et al. and tested in coal flotation as a frother. Rhamnolipid was subsequently discovered to be a promising frother or co-frother in mineral processing techniques by Khoshdast et al. Additionally, Chinese researchers' flotation desulfurization trials on high-sulfur coal with surface pretreatment using *Rhodopseudomonas spheroides* and *Thiobacillus ferrooxidans* have produced positive results.

Surfactants can be divided into anionic, cationic, and nonionic surfactants based on the polar group's dissociation characteristics. It was discovered that the impact of surfactants on the viscosity and emulsification property was widely explored. By including the right surfactant in the biodesulfurization process, you may increase the contact between bacteria and coal grain, because the process will be more effective.

The coal from the Liupanshui mine served as the experimental object in this experiment. To desulfurize the coal, a mixed culture of the bacteria *Acidithiobacillus caldus* and *Acidithiobacillus thiooxidans* was used, along with three different types of surfactants: anionic (SDS), cationic (DTAB), and nonionic (Tween 20). The three surfactants' impact on the desulfurization system was shown in this setting.

9.7 Economic Feasibility

9.7.1 Fields of Developments

Most people believe that biological and biotechnological processes are expensive and designed for the manufacture of high-value, low-volume goods. BDS requires water and is a long procedure. The size of the two-phase reactor needed to achieve the necessary desulfurization on an industrial scale is unfathomable. Therefore, it is unquestionably preferable to use a bioprocess that consumes a lower aqueous/hydrocarbon phase ratio. Enzymes need less water to function than bacteria do. However, the use of complete cells as biocatalysts is more realistic because many enzymes, NADH, and FADH2 are involved in BDS and their activity is low. Furthermore, compared to anaerobic bacteria, which can produce 0.1 g/g DCW/h, and aerobic microbes, which can produce 0.4 g/g DCW/h, cell-free extracts have lesser activity, in the range of 0.01 g of DBT removed/g protein/h. The hydrocarbon variety is another thing to take into account. There are numerous fractions used in refineries that contain organic sulfur. Similar to other microbial processes, microbial desulfurization is sluggish and requires careful monitoring to produce repeatable results. This process involves either growing microorganisms in an aqueous/oil two-phase system or desulfurizing resting cells. Growing living microorganisms in two-phase (aqueous/oil) systems is not feasible due to the lengthy process and low desulfurization yield. When creating resting cells using a high cell density method, it's possible for other bacteria that can attack C-C bonds or ingest nutrients and develop more quickly to become cross-contaminated. Additionally, large-scale facilities are needed for resting cell generation in order to prepare cells and preserve biocatalyst activity while they are transported to the desulfurization location. From a process standpoint, biodesulfurization entails the preparation of feedstock, the development of microorganisms and biocatalysts, desulfurization in a bioreactor, and separation and recovery. The development of biocatalysts with a long half-life and high, repeatable specific activity is crucial among the previously mentioned criteria. The quest for a strain with higher BDS activity and broad adaptability to various OSCs and the development of a recombinant biocatalyst with a stable activity to work in combination with refining speed represent the two biggest challenges to approaching industry-level BDS to date. In addition, greater effort is needed to increase the driving force from one phase to the next and prevent the accumulation of inhibitors in order to achieve higher desulfurization specific activity. The removal of the cooling period needed after HDS is one of the factors that could speed up BDS. Thus, one of the key elements is to develop solvent-tolerant, thermophilic biodesulfurizing bacteria with broad adaptability to various OSCs. Using nanotechnology to prepare the biocatalyst and use it in reactors is another significant, cost-effective strategy. *Pseudomonas stutzeri* immobilized in alginate was reported to have a prolonged cell survival of 600 h by reactivation, but *R. erythropolis* KA2–51 immobilized in photo-crosslinkable resin prepolymer (ENT-400 and ENTP-4000) had a prolonged cell viability of 900 h. Therefore, further research is required into the desulfurization, denitrogenation, and heavy metal removal of heavy crude oil using immobilized bacteria.

9.8 Conclusion

The two key steps for the successful commercialization of BDS are (1) maintaining a high rate of technical advancement to make it competitive with chemical desulfurization processes and for its implementation in realistic assessments pilot-plant studies and (2) figuring out the best ways to incorporate biotechnology into refineries, which primarily rely on the chemical processes and have also made significant technological advancements. HDS is still a technically and economically viable unit, especially in light of the development of new chemical catalysts that are not alumina-based. This is true even when the consumption of hydrogen and catalyst rises. However, a high inhibitory three and bigger ring aromatic hydrocarbon concentrations fundamentally control the desulfurization chemistry of heavy oils. Therefore, BDS for substantially replaced DBTs may regain some interest. A refinery's current processes must be significantly altered in order to integrate a BDS process. In order to avoid interfering with regular refining operations, the BDS process must run as quickly and dependably as other refinery processes. Therefore, despite the high level of interest in and promise for BDS, it is still difficult to advance it to a point where it can be used in refineries. However, BDS biocatalysts (whole cells or enzymes) can also be employed to detoxify chemical warfare substances or to produce other beneficial products like surfactants, antibiotics, polythioesters, and different specialized chemicals in addition to BDS of petroleum and its fractions and derivatives. However, the interdisciplinary involvement of experts in biotechnology, biochemistry, refining processes, and engineering is crucial in order to develop BDS as a supplementary process.

References

1. Jatoi AS, Aziz S, Soomro SA (2021) Effect of native microorganism Rhodococcus spp. SL-9 for dibenzothiophene degradation and its application towards clean coal approach. Cleaner Eng Technol 3:100126
2. Jatoi AS, Aziz S, Soomro SA (2021) Experimental and numerical investigation of DBT degradation via Rhodococcus spp. (SL-9) through the use of biological assisted method. Biomass Convers Biorefinery 13:1–9
3. Jatoi AS, Hashmi Z, Anjum A, Bhatti ZA, Siyal SH, Mazari S et al (2021) Overview of bioelectrochemical approaches for sulfur reduction: current and future perspectives. Biomass Convers Biorefinery 13:1–16
4. Jatoi AS, Aziz S, Soomro SA (2024) Bioprocess development and optimization of parameters for enhancing bio-assisted method for sulfur removal from Lakhra coal using Rhodococcus sp. SL-9. Int J Coal Prep Util 44:37–50
5. Jatoi AS, Aziz S, Soomro SA (2021) Biological assisted organic sulfur removal from low rank indigenous coal using airlift bioreactor. Bioprocess Biosyst Eng 44:417–427
6. Usto MA, Abbasi SA, Maitlo G, Iqbal A, Jatoi AS, Ahmed S et al (2022) Desulfurization of thar lignite by oxidative alkali leaching under pressure. Int J Coal Prep Util 42:3430–3450
7. Çelik PA, Aksoy DÖ, Koca S, Koca H, Çabuk A (2019) The approach of biodesulfurization for clean coal technologies: a review. Int J Environ Sci Technol 16:2115–2132
8. Aksoy DO, Aytar P, Toptaş Y, Çabuk A, Koca S, Koca H (2014) Physical and physicochemical cleaning of lignite and the effect of cleaning on biodesulfurization. Fuel 132:158–164
9. Chen S, Zhao C, Liu Q, Zang M, Liu C, Zhang Y (2018) Thermophilic biodesulfurization and its application in oil desulfurization. Appl Microbiol Biotechnol 102:9089–9103
10. Etemadi N, Sepahy AA, Mohebali G, Yazdian F, Omidi M (2018) Enhancement of biodesulfurization capability of a newly isolated thermophilic bacterium using starch/iron nanoparticles in a controlled system. Int J Biol Macromol 120:1801–1809
11. Kotelnikov V, Saryglar CA, Chysyma R (2020) Microorganisms in coal desulfurization. Appl Biochem Microbiol 56:521–525
12. Konishi J, Ishii Y, Onaka T, Okumura K, Suzuki M (1997) Thermophilic carbon-sulfur-bond-targeted biodesulfurization. Appl Environ Microbiol 63:3164–3169
13. Ishii Y, Konishi J, Okada H, Hirasawa K, Onaka T, Suzuki M (2000) Operon structure and functional analysis of the genes encoding thermophilic desulfurizing enzymes of Paenibacillus sp. A11-2. Biochem Biophys Res Commun 270:81–88
14. Becker C, Marder M, Junges E, Konrad O (2022) Technologies for biogas desulfurization-an overview of recent studies. Renew Sust Energ Rev 159:112205
15. Kim TS, Kim HY, Kim BH (1990) Petroleum desulfurization by Desulfovibrio desulfuricans M6 using electrochemically supplied reducing equivalent. Biotechnol Lett 12:757–760
16. Tilstra L, Eng G, Olson GJ, Wang FW (1992) Reduction of Sulphur from polysulphidic model compounds by the hyperthermophilic Archaebacterium Pyrococcus furiosus. Fuel 71:779–783
17. Finnerty WR (1993) Organic sulfur biodesulfurization in non-aqueous media. Fuel 72:1631–1634

18. Xu J, Liu X, Song C, Du Z, Wang F, Luo J et al (2020) Biodesulfurization of high sulfur coal from Shanxi: optimization of the desulfurization parameters of three kinds of bacteria. Energy Sour Part A Recov Util Environ Effects 42:2297–2315
19. Hokmabadi M, Khosravinia S, Mahdavi MA, Gheshlaghi R (2022) Enhancing the biodesulphurization capacity of Rhodococcus sp. FUM94 in a biphasic system through optimization of operational factors. J Appl Microbiol 132:3461–3475
20. Marcelis C (2002) Anaerobic biodesulfurization of thiophenes. Wageningen University and Research
21. Maslova O, Senko O, Akopyan A, Lysenko S, Anisimov A, Efremenko E (2021) Nanocatalysts for oxidative desulfurization of liquid fuel: modern solutions and the perspectives of application in hybrid chemical-biocatalytic processes. Catalysts 11:1131
22. Martínez I, Mohamed ME-S, Santos VE, García JL, García-Ochoa F, Díaz E (2017) Metabolic and process engineering for biodesulfurization in gram-negative bacteria. J Biotechnol 262:47–55
23. Dejaloud A, Vahabzadeh F, Habibi A (2017) Ralstonia eutropha as a biocatalyst for desulfurization of dibenzothiophene. Bioprocess Biosyst Eng 40:969–980
24. Boshagh F, Rahmani M, Rostami K, Yousefifar M (2021) Key factors affecting the development of oxidative desulfurization of liquid fuels: a critical review. Energy Fuel 36:98–132
25. Kalita M, Chutia M, Jha DK, Subrahmanyam G (2022) Mechanistic understanding of Gordonia sp. in biodesulfurization of organosulfur compounds. Curr Microbiol 79:82
26. Adlakha J, Singh P, Ram SK, Kumar M, Singh M, Singh D et al (2016) Optimization of conditions for deep desulfurization of heavy crude oil and hydrodesulfurized diesel by Gordonia sp. IITR100. Fuel 184:761–769
27. Zargar AN, Mishra S, Kumar M, Srivastava P (2022) Isolation and chemical characterization of the biosurfactant produced by Gordonia sp. IITR100. PLoS One 17:e0264202
28. Esfahanipour A, Jahanbin R (2017) An application of genetic network programming model for pricing of basket default swaps (bds). AUT J Model Simul 49:13–22
29. Rashidi L, Dariani JT, Khosravi-Darani K (2007) Biodesulfurization: biochemical and genetic engineering aspects. Energy 50:1992–1998
30. Mohebali G, Ball AS (2008) Biocatalytic desulfurization (BDS) of petrodiesel fuels. Microbiology 154:2169–2183
31. Martínez I, Mohamed ME-S, Rozas D, García JL, Díaz E (2016) Engineering synthetic bacterial consortia for enhanced desulfurization and revalorization of oil sulfur compounds. Metab Eng 35:46–54
32. Khan J, Ali MI, Jamal A, Achakzai JK, Shirazi JH, Haleem A (2023) Assessment of the dibenzothiophene desulfurization potential of indigenously isolated bacterial consortium IQMJ-5: a different approach to safeguard the environment. Arch Microbiol 205:95
33. He Z, Xiao H, Tang L, Min H, Lu Z (2013) Biodegradation of di-n-butyl phthalate by a stable bacterial consortium, HD-1, enriched from activated sludge. Bioresour Technol 128:526–532
34. Martinez I, Santos VE, Gomez E, Garcia-Ochoa F (2016) Biodesulfurization of dibenzothiophene by resting cells of pseudomonas putida CECT5279: influence of the oxygen transfer rate in the scale-up from shaken flask to stirred tank reactor. J Chem Technol Biotechnol 91:184–189
35. Martínez I, Mohamed ME-S, García JL, Díaz E (2022) Enhancing biodesulfurization by engineering a synthetic dibenzothiophene mineralization pathway. Front Microbiol 13:987084

MIX
Papier aus verantwortungsvollen Quellen
Paper from responsible sources
FSC® C105338

If you have any concerns about our products,
you can contact us on
ProductSafety@springernature.com

In case Publisher is established outside the EU,
the EU authorized representative is:
Springer Nature Customer Service Center GmbH
Europaplatz 3, 69115 Heidelberg, Germany

Printed by Libri Plureos GmbH
in Hamburg, Germany